Manual of Water Supply Practices

M71

Climate Action Plans—Adaptive Management Strategies for Utilities

American Water Works Association

Manual of Water Supply Practices—M71

Climate Action Plans—
Adaptive Management Strategies for Utilities

If you find errors in this manual, please email books@awwa.org.
Possible errata will be posted at www. awwa.org/M71.

Managing Editor/Project Manager: Melissa Valentine
Technical Editor: Suzanne Snyder
Cover Design/Technical Illustrations: Mike Labruyere
Manager, Publishing Operations: Gillian Wink
Production: Innodata
Senior Specialist - Manuals: Willadee Hitchcock
Manuals Coordinator 2: Janet Greifinger

Library of Congress Cataloging-in-Publication Data

Names: Tummuri, Spandana, author. | American Water Works Association, issuing body
Title: M71 - climate action plan / by Spandana Tummuri.
Other titles: Climate action plan
Description: Denver, CO : American Water Works Association, 2021. | Includes bibliographical references and index. | Summary: "The manual aims to help small and mid-sized utilities develop climate action plans to address adaptation and mitigation strategies countering the impact of climate change. The ultimate goal of the manual is to provide a starting point for utility managers, if climate change adaptation planning and developing climate action plans is the topic of their interest"-- Provided by publisher.
Identifiers: LCCN 2020020759 | ISBN 9781625763693 (paperback)
Subjects: LCSH: Climate change mitigation. | Water utilities--Management.
Classification: LCC TD171.75 .T85 2020 | DDC 363.738/747--dc23
LC record available at https://lccn.loc.gov/2020020759

Printed in the United States of America

ISBN 978-1-62576-369-3 ISBN, electronic 978-1-61300-537-8

DOI https://doi.org/10.12999/AWWA.M71ed1

American Water Works Association

American Water Works Association
6666 West Quincy Avenue
Denver, CO 80235-3098
awwa.org

Contents

Figures

Tables

This page intentionally blank.

Preface

American Water Works Association (AWWA) created the Climate Change Committee under the Water Resources Sustainability Division to prepare its members for the "new normal" scenario, in which the potential effects of climate change must be considered for long-term sustainability of our local, regional, national, and global water resources. This manual is geared toward achieving the Climate Change Committee's goal to document the status of knowledge and describe current and future research activities that will enhance utilities' understanding of climate change effects and mitigation/adaptation techniques. Specifically, the manual aims to help small and mid-sized utilities develop climate action plans (CAPs) to address adaptation and mitigation strategies countering the impact of climate change.

The committee recognizes that those effects may be more or less severe depending on each utility's location and situation, but what the committee is recommending is that each utility assess its vulnerability to climate change and prioritize mitigation and adaptation measures in the context of other risks to their operations.

The manual is organized in a way to introduce the concept of climate change to readers by providing the fundamentals of climate change planning and decision making. The concepts of engineering design and planning in the realm of climate change uncertainty is discussed. The manual also includes a detailed summary of the various impacts of climate change, specifically from a utility perspective, and provides the step-by-step process of developing a CAP. The manual includes a brief summarization of the challenges and issues associated with the public and stakeholder engagement. A review of various CAP case studies is included as part of the manual to familiarize readers with the other research material available for their use. In a world of climate change planning, where information and science is changing and improving at a breakneck speed, the authors' goal is to provide relevant reference material for the information presented. The manual will be reviewed and revised as and when information presented in the manual needs to be updated.

The ultimate goal of the manual is to provide a starting point for utility managers for whom climate change adaptation planning and developing a CAP are of interest. While the manual does not lend the reader to preparing a fully developed CAP, it guides the reader to all the relevant information they will need to develop one on their own. There is no cookie-cutter approach that can be applied to the process of developing climate action plans that includes a step-by-step checklist as the infrastructure vulnerabilities, the climate change impacts, the stakeholder engagement, and the adaptation and response measures vary by region and utility.

The template for the manual is set in such a way that the reader can easily access any of the chapters individually, without having to read the entire manual from the beginning to the end, to grasp the content in the individual chapters. Each chapter is focused on a specific topic and the relevant information supporting the chapter material and the references are included along with the chapter content.

This page intentionally blank.

Acknowledgments

The authors would like to thank American Water Works Association (AWWA) for providing the staff support in the successful oversight and completion of this manual. Several AWWA staff members have played a key role in the development of this manual, including Molly Beach, Janet Greifinger, Sue Weikel, Willadee Hitchcock, Julie Kurzawa, Alex Gerling, Peggy Tyler, Adam Carpenter, Amy Brown, and Dawn Flancher.

This manual would not have been possible without the continued encouragement and support from the leadership of AWWA, the Technical and Educational Council (TEC) leadership, the Water Resource Sustainability Division (WRSD) leadership, and the Climate Change Committee (CCC) leadership and membership.

Our thanks to the Technical and Educational Council committee, the Steering Committee, and the WRSD (Jonathan Keck, John Hughes, Laurens Van der Tak, Brent Burton, and Bill Davis) for their respective reviews and timely balloting of the approval for the manual development.

Special thanks to Chi Ho Sham (former chair of WRSD) for his continued support in the manual development process, Brent Burton (former liaison for Climate Change Committee), Laurens Van der Tak (current chair of WRSD), Sarah Deslauriers (former chair of Climate Change Committee), Bill Davis (former chair of WRSD), Jennifer Heyman (current liaison for Climate Change Committee), Michelle Maddaus (former chair of Climate Change Committee).

Our sincere thanks goes to the staff and leadership of the US Environmental Protection Agency (USEPA) for its continued response to our questions related to USEPA's Creating Resilient Water Utilities (CRWU) initiative and the Climate Resilience Evaluation and Awareness Tool (CREAT).

Our thanks to the manual review committee members who did a thorough review of the chapters and provided valuable feedback to improve the quality of the manual content. The list of reviewers includes Adam Carpenter, Barney Austin, Bruce Rhodes, Dan Haddock, Jacqueline Koch, Jack Moyer, James Chelius, Jim Miller, Laurna Kaatz, Michael McMahon, Pat Davis, Paul Granger, Paul Selsky, Scott Weikert, and Tom Gooch.

Our special thanks is due to Ms. Destree Lazo Bascos, Roger Rempel, Zachary Rybarczyk, and Steve Conrad owing to their valuable contributions to the manual development process.

This manual was completed owing to the countless volunteer hours spent by this team of diligent and tireless authors. Their spirit, expertise, and leadership made this manual possible. The list of authors includes Adam Carpenter, Andrew Schwarz, Brent Burton, Catherine Smith, Eme Williams-Blake, Enrique Lopezcalva, Jack Moyer, James Chelius, Kavita Heyn, Laurens Van der Tak, Michelle Maddaus, Nirvana Ward, Rajat Chakraborti, Sarah Deslauriers, Spandana Tummuri, and Tirusew Asefa.

This page intentionally blank.

Executive Summary

American Water Works Association (AWWA) created the Climate Change Committee under the Water Resources Sustainability Division to prepare its members for the "new normal" scenario, in which the potential effects of climate change must be considered for long-term sustainability of our local, regional, national, and global water resources. The manual aims to help small and mid-sized utilities develop climate action plans to address adaptation and mitigation strategies countering the impact of climate change. The ultimate goal of the manual is to provide a starting point for utility managers for whom climate change adaptation planning and developing climate action plans are of interest. While the manual itself doesn't lend the reader to preparing a fully developed climate action plan, it guides the reader to all the relevant information so that they able to develop a report on their own.

The manual is divided into eleven chapters, each chapter addressing various aspects of the climate action plan development process. Together, the 11 chapters form a cohesive source of information that guides utility managers through the process of developing climate action plans.

Chapter 1—Introduction includes discussion on the importance of climate change impacts on various aspects of water resource management and utility infrastructure planning and management. Developing a sound climate action plan (CAP) that addresses the adaptation and mitigation aspects of site-specific climate change impacts is critical to efficient management of water utilities. The purpose of this manual is to provide a comprehensive guidance document to generate, review, report, and disseminate information on water supply planning and management relating to issues of climate change, both mitigation and adaptation, and to facilitate awareness among water utilities and the broader community on developing and maintaining CAPs. Impacts of climate change on water quality, quantity, operational reliability, and financial and institutional aspects of utility management are discussed in this chapter. The various solutions to help address the impacts that are discussed include mitigation measures, long-term planning processes, demand management, operational controls, carbon footprint reduction, energy management, regional cooperation, building resiliency by securing additional water supplies, encouraging public participation by raising awareness, and other management measures.

Chapter 2—Purpose and Audience is a key chapter that sets the foundation for the relevance and application of this manual. It describes the purpose of the manual and its intended audience. So much information is available in the climate change scientific community in the form of reports funded by public and private entities. It is important to discern what information is relevant and what is not. Along similar lines, it is important to clarify what the stated goals of this manual are and who benefits from the use of this manual.

Chapter 3—Science of Climate Change is perhaps one of the most important chapters in the entire manual. Before venturing into the process of understanding and developing CAPs, it is important to have a basic understanding of the science of climate change. There are thousands of peer-reviewed and published reports that discuss the science of climate change. The reader could very easily get lost in that knowledge. For readers who are initially trying to grasp the details of climate change science, it is easy to get confused about the level of detail to focus on. In this chapter, the science of climate change and the important pieces of information required to develop climate action plans is gathered and presented to the reader in a concise manner. The information is also presented at a level of detail that is appropriate for utility managers and utility operators. The chapter includes a discussion on the details of climate change variability, greenhouse gas emissions and how they are used to measure and monitor climate change, and global climate model projections and inherent assumptions. The chapter also includes a summary of the impact of climate change on the hydrologic cycle and how utility managers should plan for a future that is driven by climate uncertainty. Several resources are presented and discussed in detail for readers who wish to gather more information on certain portions of the information presented in the chapter.

Regional impacts of climate change are discussed in this chapter. It is well known that climate change does not impact all regions and hydrologic environments in the same manner. Therefore, a cookie-cutter approach to analyzing climate change science and developing adaptation and mitigation measures does not serve well for utility management. It is important to understand the local and regional impacts of global climate change so that adaptation and mitigation measures discussed in the CAP are geared toward directly addressing the relevant impacts. Finally, the chapter provides a summary of the system vulnerability and various methods for assessing vulnerability.

Chapter 4—Principles of Engineering Practice is a utility managers' guide for operating and managing infrastructure while accounting for the effects of climate change. All utility managers are trained in the traditional concepts of engineering practices that include evaluating past practice, designing for a predictable future, defining level of service for infrastructure, assuming redundancy and resiliency in the form of factor of safety, and traditional engineering approaches to problem solving. However, these traditional concepts are challenged by the uncertainty associated with climate change impacts. Most engineering assumptions and practices that served utility managers in the past can fail when trying to plan for an uncertain future. This chapter includes a detailed discussion on various approaches adopted by utilities all over the world to incorporate the uncertainty into their planning and engineering processes. The chapter includes a discussion on climate change impact assessment and lays out best management practices for arriving at design solutions. Practical planning strategies and adjustments to defining level of service, planning for new infrastructure, and the role of engineering judgment are discussed.

Chapter 5—What Is a Climate Action Plan? defines exactly what a CAP is and is not. Most of the information presented in the previous chapters is relevant to understand climate change and determine how to apply engineering practice to address climate change impacts. Before the reader embarks upon the journey of developing the CAP, a clear description is required to understand what a cap actually is. The chapter tries to define the impacts that the climate action plan must address and the measure and actions it must include. The chapter also includes a discussion on the tools available to develop and prepare the climate action plans. These tools act as guidance material only and the ultimate responsibility of developing a climate action plan tailored to the utility needs falls on the shoulders of the utility managers.

Chapter 6—Why Do Utilities Need a Climate Action Plan? focuses on a relevant and important question. While it is good to understand the science and pertinent impacts of climate change, utility managers need to ask themselves these questions: *Do we need to develop a climate action plan and if so, to what end? What is the primary purpose of developing a climate action plan and how do we plan to integrate the details of the climate action plan into day-to-day utility operations?* The chapter discusses the key drivers for developing CAPs and the details of evaluating the vulnerability and resiliency of the utility in addressing climate change impacts. The information in this chapter is presented in a practical manner with the help of various examples from events and scenarios driving the development of CAPs for various entities in the United States.

Chapter 7—How Do Utilities Develop a Climate Action Plan and Keep It Current? This key chapter of the manual includes a detailed discussion on how utilities actually develop the CAP. The reader will have enough background information from Chapters 1–6 to understand what it takes to prepare a CAP. By following the information presented in this chapter, the reader can effectively develop a CAP that serves the utility's needs in the best possible manner. The chapter includes a plan for developing the mitigation and adaptation aspects of the CAP. The discussion about addressing the adaptation measures in the CAP includes assessing vulnerabilities, setting a context for the climate action plan development, selecting adaptation strategies and actions and implementing the climate action plan. Most of the discussion in this chapter is focused on the adaptation measures that are typically included in a CAP as that is the primary focus of the manual. The chapter does discuss mitigation measures but in less detail.

Chapter 8—Stakeholder Involvement and Public Participation is the most important aspect of successful development and implementation of the CAP. The chapter focuses on identifying the key stakeholders, defining the terms and conditions for public participation, encouraging and involving public feedback in the planning process, and developing a detailed plan for communicating and seeking feedback on various plan milestones. Since stakeholders and the public tend to be the consumers, customers, and rate payers of the water utility, it is important to engage them from the beginning of the process.

Chapter 9—Estimating the Cost of a Climate Action Plan is a brief summary of the process of determining the cost associated with the development and implementation of CAPs. Most utilities benefit significantly if appropriate funding sources are available to fund various infrastructure projects. The chapter provides detailed information on the funding sources available for undertaking the process of developing a CAP from the ground up. The chapter also includes a summary of case studies of utilities where CAPs were developed and details of the costs incurred by those entities.

Chapter 10—Case Studies of Climate Action Plans and Adaptation Measures is the most relevant chapter for utilities and readers interested in developing CAPs. Before developing this manual, the authors of spent a significant amount of time conducting gap analysis and research on the available information related to CAPs. While reviewing the information available, the authors compared the information readily available with the intended purpose of this manual to avoid any redundancies and replication of efforts. The summary of the review process and the information collected in the process of reviewing the case studies is included in the chapter. While the case studies cannot be directly adopted and used by a specific utility, they can serve as reference material and guidance documents while developing a CAP catered to meeting the needs of a specific utility.

Chapter 11—Summary and Conclusions includes a brief discussion of why this manual is important and how this work will be beneficial to water resource managers and utilities in general. There is no bigger concern for utility managers than that of not providing the services intended to be provided by the utility. While many factors impact the ability of a utility to serve their customer needs, no other factor impacts the utility operations in a complex, uncertain, and overwhelming way like climate change does. Already, there are several instances of utilities facing the worst-case impacts driven by climate change and being caught unprepared and off-guard in terms of responding to the impacts of climate change. The impacts of various hurricanes on the East and Southeast coasts of the United States are a very recent example. The staggering impact of the wildfires in California is another example of the overwhelming scale and magnitude of climate change. In the face of uncertainty and the exponential increase in the frequency of extreme events, it seems prudent for utilities to understand the impacts of climate change and their system vulnerabilities and develop a CAP to stay prepared and resilient.

Chapter 1

Introduction

Addressing the impacts due to climate change on water utilities is a challenge faced by water utility managers on a day-to-day basis. The varied types of impacts due to climate change combined with the variety of system vulnerabilities within a water utility pose challenges for water utility managers. Developing a sound climate action plan (CAP) that addresses the adaptation and mitigation aspects of site-specific climate change impacts is critical to efficient management of water utilities. The purpose of this manual is to provide a comprehensive guidance document to share information on water supply planning and management relating to issues of climate change, both mitigation and adaptation, and to facilitate awareness among water utilities and the broader community on developing and maintaining climate action plans.

The manual provides step-by-step guidance to water utility managers by helping them define the various concerns associated with the impacts of climate change to management of water supplies. Chapter 1 includes a comprehensive summary of the various threats to water utility management due to climate change and the severity of the impact. Chapter 2 includes a brief review on the purpose of the document and its intended audience. Chapter 2 also includes a description of frequently used terms to ensure uniform applicability and understanding of some of the frequently used terms in the field of climate change and water resource management.

Chapter 3 is a detailed review of the science of climate change and it includes a comprehensive summary of the "what," "why," and "how" of the climate change dynamics. Chapter 4 connects the science of climate change to the engineering and describes the importance of incorporating the study of impacts of climate change for efficient management of water utilities and water resources. The most elegant way of incorporating the study of climate change impacts into utility management is to develop a CAP. What is a CAP and what constitutes the process of developing a CAP is described in Chapter 5. Chapter 6 makes a very strong case for why developing a CAP is important for effective management of water supplies and water infrastructure. It also talks about the challenges and issues associated with not developing the CAP. Chapter 7 includes a detailed description of the process of developing the CAP and the steps necessary for preparing a meaningful CAP for any utility. Public participation and stakeholder management are major drivers for developing and implementing a CAP. Chapter 8 covers the details of addressing the issues associated with public and stakeholder participation along with a summary of techniques for improving this approach. Chapter 9 provides a brief summary of the approach for costing the CAPs and an estimate of the effort and budget involved in preparing the CAP. Chapter 10 is a compilation of case studies of CAPs and other resources that are helpful to water utility managers when faced with the challenge of developing a CAP for their utility.

UNDERSTANDING THAT CLIMATE CHANGE IS A REALITY

After decades of debate, most scientists and engineers throughout the world have generally accepted climate change as a reality. A report from the United Nations Intergovernmental Panel on Climate Change (IPCC) was released in 2007 (IPCC 2007). Written by a panel of senior scientists from around the world, the report finds the proof of climate change "unequivocal."

The findings documented in the 2007 IPCC report and the updated 2014 IPCC report substantiate global climate change and conclude that human activities have accelerated most of these observed changes during the past 50 years. The key conclusions of the report are as follows.

1. Warming of the climate system is evidenced by increased global average air and ocean temperatures, increase in greenhouse gases (GHGs), and rising sea level. These changes recorded during the past 85–150 years are documented in the IPCC Report and are shown in Figure 1-1.
2. During the past 100 years, the Earth has warmed by 1.5°F (0.76°C) on average, and the rate of warming has increased over time. The 11 warmest years on record have all occurred in the past 12 years. The latest data show the warmest 17 years on record have all occurred since 2001. (NASA, nd)
3. Rates of observed sea level rise have almost doubled in the past 50 years, increasing from 18 cm per century during 1961–2003 to 31 cm per century during 1993–2003. The concentration of carbon dioxide (CO_2) and other GHGs in the atmosphere have continued to increase because of man-made emissions. Concentrations of CO_2 and methane are at their highest levels in at least 650,000 years.
4. Extreme weather events have increased, and regional climate patterns are changing. Heat waves and other weather extremes, as well as changes in atmospheric circulation patterns, storm tracks, and precipitation, can now be traced back to climate change.

Past climate variability shows that significantly different climate conditions occurred just before the standard hydrologic record, as demonstrated by paleoclimate reconstructions from tree rings or other methods used to understand climate history. Some agencies have begun to use stream flow reconstructions based on tree-ring analysis, and papers have been written to support using this method in decision-making (Rice et al. 2009).

Though the evidence proves the existence of climate change, the entire gamut of effects from such change is not yet fully understood and is difficult to quantify. Figure 1-2 shows a representation of probable future effects of climate change on our water environment. The effects of climate change on our water resources could significantly change the way our resources are managed and how water utility business is conducted.

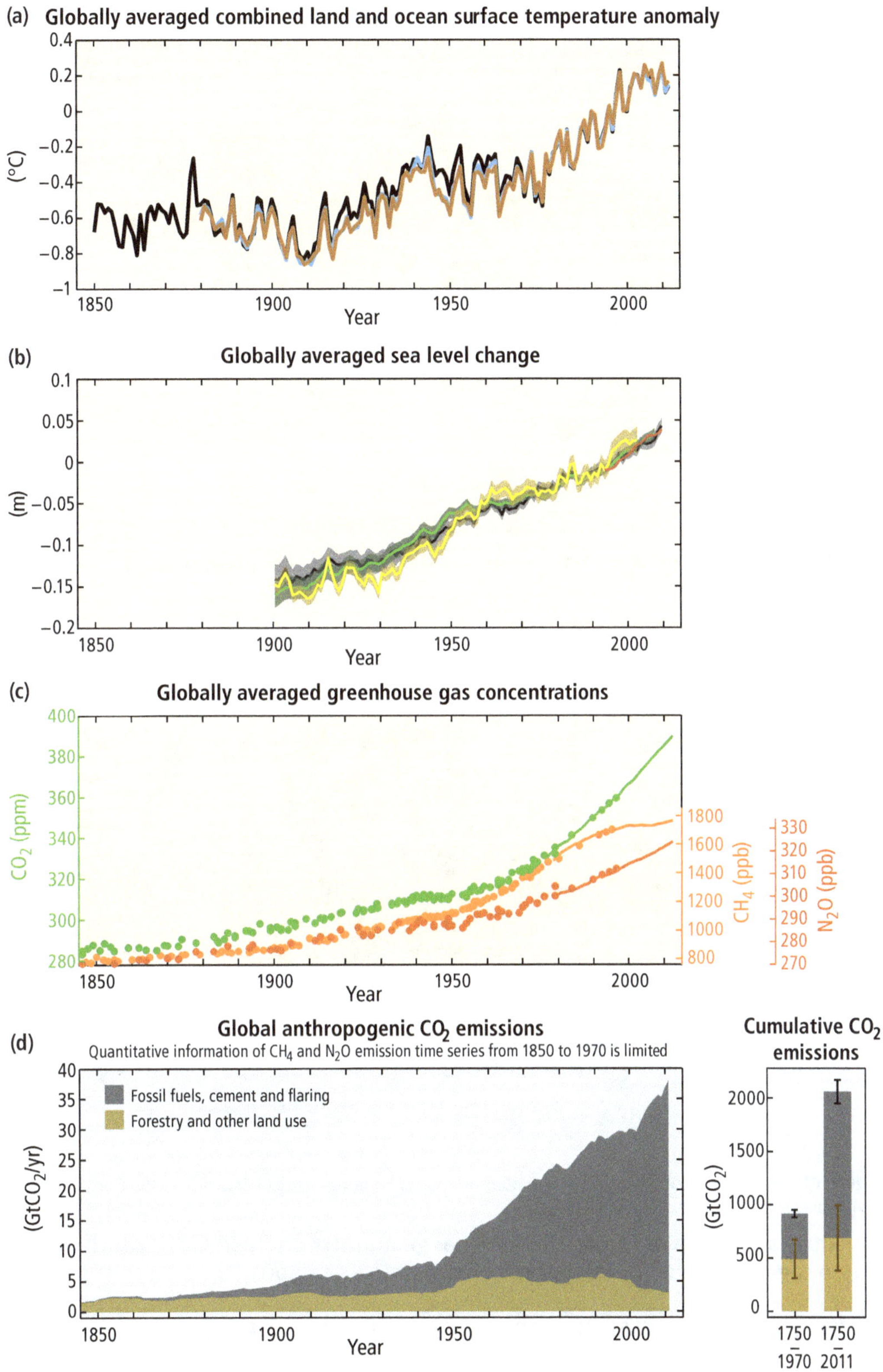

Source: IPCC, 2014: Climate Change 2014: Synthesis Report. Contribution of Working Groups I, II and III to the Fifth Assessment Report of the Intergovernmental Panel on Climate Change

Figure 1-1 Changes in global average air and ocean temperatures, greenhouse gases (GHGs), and sea level during the past 85–150 years

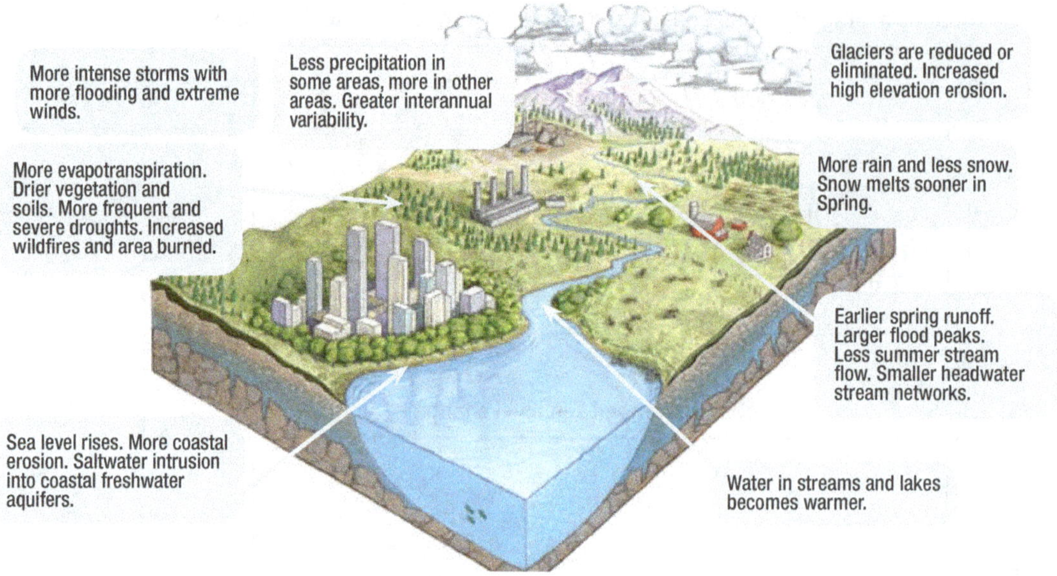

Source: US Department of Agriculture, FS 908 - Sustaining Healthy Watersheds, June 2008

Figure 1-2 A representation of probable future effects of climate change on our water environment

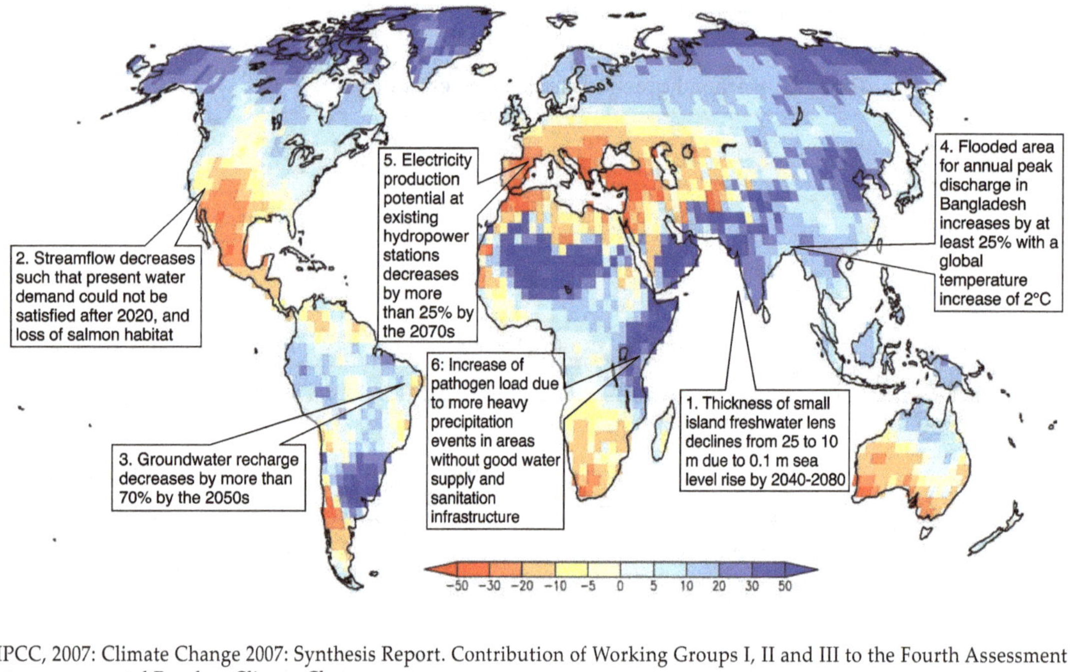

Source: IPCC, 2007: Climate Change 2007: Synthesis Report. Contribution of Working Groups I, II and III to the Fourth Assessment Report of the Intergovernmental Panel on Climate Change

Figure 1-3 Illustrative map of future climate change impacts on freshwater which are a threat to the sustainable development of the affected regions

APPRECIATING THE POTENTIAL EFFECTS OF CLIMATE CHANGE ON WATER UTILITIES

Scientists and engineers continue to work to define the full scope of effects and timing of climate change on our water resources, which will affect the quantity and quality of water available for potable use, food production, industrial use, power generation, and other vital services. Figure 1-3 shows some predicted climate change effects related to freshwater availability in various regions of the world.

A report prepared by CH2M HILL and the Association of Metropolitan Water Agencies (AMWA) quantified the costs of adaptation for water utilities through 2050 under various climate change scenarios, concluding the costs to be approximately $325 billion to $692 billion (CH2M HILL 2009). The Water Research Foundation (WRF), in its research, has broadly categorized climate change effects on water resources as water quantity effects, water quality effects, operational reliability effects, and financial and institutional effects.

Water Quantity Effects

The largest effects of climate change on water resources include changes in temperature and precipitation variability; some areas may receive more precipitation and others much less. These changes are reflected in greatly increased or reduced stream flows, changes in timing and intensity of precipitation, changes in snowmelt quantity and timing of runoff, and changes in aquifer recharge.

Balancing increasing water demand associated with warmer temperatures with reductions in available supply associated with decreased precipitation will be particularly challenging for some suppliers. Unpredictability related to the uncertainty of the types and temporal nature of effects and other complex water resources challenges (e.g., Endangered Species Act, Clean Water Act, Safe Drinking Water Act compliance) are likely to become significant management issues for utilities.

Water Quality Effects

Climate change effects such as increased precipitation intensity could cause changes in sediment and pathogen loading in urban runoff and increase sewer overflows. Temperature changes could cause algal blooms/eutrophication of surface water bodies, changes in watershed vegetation, and changes in growth/migration patterns of invasive species (e.g., quagga mussel, zebra mussel). Saltwater intrusion caused by rising sea levels could threaten the quality of groundwater supplies. Such changes in source water quality can create significant treatment, operational, and budgetary challenges for water utilities.

Operational Reliability Effects

Changes in the quantity and quality of available water supplies could affect operational reliability of water utilities. Optimization of water supplies could require increased monitoring in conjunction with other control techniques such as water conservation/demand management, recycling and reuse of water as well as the use of nonpotable supplies, and the greater uptake of non–climate-dependent supply sources such as seawater desalination. Offsetting reduced surface water supplies with increased groundwater pumping could result in lowering of the groundwater table. A lowered groundwater table could in turn cause soil drying, subsidence issues, and settling and an associated increase in pipe breakage and water loss. Saltwater intrusion will increase pipe corrosion in affected areas and may force utilities to implement expensive groundwater treatment and/or create hydrologic barriers to prevent saltwater intrusion. Growth of invasive species may cause complex operations and maintenance issues in addition to further reducing available supplies by outcompeting native species with lower water consumption requirements (also, increased ambient temperatures may have implications for sewer corrosion rates and treatment processes).

Utility managers will likely need to invest more time and money in understanding the extent of possible climate risks and effects to maintain system reliability in the future. The implementation of real-time monitoring technology, along with conjunctive water resources management practices, may become vital to sustaining operations. Additionally, costs associated with protecting infrastructure from effects may be significant.

Financial and Institutional Effects

Effects on water supply quantity, quality, and operational reliability will necessitate changes in utility operations. The projected increase in operational complexity in future years may result in a significant increase in operational costs. Increased infrastructure costs associated with acquiring, conveying, treating, and distributing potable water are also expected to increase. These cost increases will generate a need for increased revenue, and utilities will likely need to develop new rates that reflect the true cost of water under these "new normal" scenarios.

Further, significant population fluctuations associated with en masse migration away from water-scarce areas may necessitate institutional changes to address changing population centers and increased competition for limited resources. Additionally, new regulatory requirements to minimize energy consumption and achieve improved per capita carbon emissions to meet greenhouse gas (GHG) mitigation goals may affect water utility operations in that investments will likely be required to implement more efficient technologies and to develop behind-the-meter renewable energy sources.

Table 1-1 presents a summary of observed/possible effects to water resources from climate change. Although many studies have defined the scope of anticipated climate change effects and a variety of literature is available with investigative study results, significant uncertainty and data gaps also exist that could cause utilities to be ill-prepared to identify and address potential effect scenarios.

CLIMATE CHANGE AND SUSTAINABILITY

Although water utility operations may be sustainable under the relatively short historical record of existing climate conditions, any change to these recorded conditions should be evaluated to determine whether utility operations will continue to be sustainable under the wider variability predicted by current downscaled global climate models. Some degree of variability is inherent in any norm or statistical average. Water utilities typically account for this variability in their planning, allowing them to develop the systems needed to provide for reliable water service. However, projected changes in climate indicate that there could be significant deviations from the historical norm and statistical variability. In other words, future climate conditions are predicted in some circumstances to deviate significantly from recent historical climate conditions, rendering obsolete the analysis of recent historical climate data as a planning tool for water utilities.

A climate change assessment conducted for the State of Washington by the Climate Impact Group (CIG) at the University of Washington ran 20 different general circulation models (GCMs), which are used to forecast future trends in the norm or average climate conditions based on various atmospheric GHG concentration scenarios (CIG 2009). They found that after 50 years, each model projected a range of temperature increases, whereas 70 percent of the models projected a precipitation increase. The projected temperature increases ranged from 1.5 to 4.5°F, and the projected precipitation changes ranged from –4 to +9 percent. In addition, climate change was projected to alter historical patterns for snowpack, storm events, flooding, sea level, and water quality.

These changes will affect water supply sources and yield, water demand, water quality, and infrastructure planning. Each of these changes plays a role in the sustainability of a water supply system. Perhaps the most obvious effects are the surface water supplies. Climate effect studies have been performed for water utilities in the western United States that are dependent on snow-fed sources or are within snow-transient zones. The consistent finding is that as the temperature rises, the snow line rises and snow volume decreases. The result of this shift is that more precipitation falls as rain in the midwinter months and lower flows start earlier in the year and on average are lower during the summer.

Table 1–1 Observed and possible effects of climate change

Observed Effect	Observed/Possible Effect
Increase in atmospheric temperature	Reduction in water availability in basins fed by glaciers that are shrinking, as observed in some cities along the Andes in South America (Ames, 1998; Kaser & Osmaston, 2002)
Increase in surface water temperature	Reduction in dissolved oxygen content , mixing patterns, and self-purification capacity
	Increase in algal blooms
Sea-level rise	Salinization of coastal aquifers
Shifts in precipitation patterns	Changes in water availability caused by changes in precipitation and other related phenomena (e.g. ground water rechange, evapotranspiration)
Increase in interannual precipitation variability	Increases the difficulty of flood control and reservoir utilization during the flooding season
Increased evapotranspiration	Water availability reduction
	Salinization of water resources
	Lower ground water level
More frequent and intense extreme events	Floods affect water quality and water infrastructure integrity, and increase fluvial erosion, which introduces different kinds of pollutants to water resources.
	Droughts affect water availability and water quality

Source: Climate Change and Water, IPCC Technical Paper VI, Intergovernmental Panel on Climate Change, June 2008.

This, in effect, reduces natural available storage and thereby reduces the reliable yield of the supply system. In the Pacific Northwest, this reduction ranges from 4 to 25 percent (CIG 2009). There is also a related shift in the timing of snowmelt, with snow melting earlier in the year, resulting in increased streamflow during the winter and decreased flow during the late spring and fall. This projected trend is significant for water suppliers, who need to provide for the highest water demands during summer months while meeting instream flow requirements during drier months of the year.

Increases in winter precipitation could increase groundwater recharge but could also result in a greater proportion of precipitation running off saturated soils and not percolating into the aquifer. Further, decreased runoff during summer dry periods could result in a net annual reduction in aquifer recharge.

On the demand side of the utility picture, warmer temperatures and drier conditions in summer months, when water demand increases because of outdoor irrigation, would likely result in increased water demands. Water demand forecasts were performed for three major counties in Washington to project the effect of climate change (CIG 2009). The result of this study was that after 50 years, the average annual water demand could increase by 5–12 percent, depending on which GCM was used to forecast temperature and precipitation changes.

Although climate change effects to surface water bodies have been fairly well-studied, there has been much less research on the effect of climate change on groundwater. This is partly attributable to the greater difficulty in measuring groundwater systems. Increases in winter precipitation would, in principle, increase groundwater recharge. However, this increase may be offset by an increase in evaporation associated with warmer temperatures. Further, an increase in precipitation intensity may result in a greater proportion of precipitation running off saturated soils rather than percolating directly to the groundwater basin. These trends, however, are dependent on the geology and soils of the specific watersheds. It is believed that natural effects to groundwater are not as well understood and not as immediate in comparison to surface water effects; however, the direct effects on groundwater are not well-substantiated by research, and effects to major aquifers may be significant. For example, one might consider the relationship between aquifers and surface waters, particularly streams and rivers, which can reduce flows if aquifers are stressed sufficiently.

Climate change effects could also affect the sustainability of some aspects of water supply infrastructure. The most affected utilities would be coastal utilities and those vulnerable to flooding. With an increase in winter precipitation falling as rain, as opposed to snow, coupled with a probable increase in precipitation during the winter months, there will be an increase in flooding risk. Water intake facilities on rivers and facilities in floodplains will be more susceptible to flood damage. To prevent damage, these facilities may have to be armored or possibly relocated. Further, water and wastewater utility infrastructure located in low-lying coastal areas could be affected by rising sea levels and saltwater intrusion, which could result in effects ranging from an increase in corrosion of buried infrastructure to complete inundation.

Water quality is expected to be adversely affected by climate change. More intense and frequent precipitation events may result in increased erosion which could lead to increased turbidity and organic material content in surface waters. This may result in elevated disinfection by-product concentrations in finished water as well as changing filter run times and operating protocols. Also, elevated temperatures can accelerate organic growth rates within water impoundments, thereby lowering filtering rates. However, higher temperatures can be beneficial as they increase filter rates (or run times) in conventional and membrane filters. In addition, higher water temperatures increase chemical reactions. So there might be some benefit in higher temperatures, including better disinfection and possibly less need for chemicals. However, this will likely be offset by more algal growth, which can decrease filter rates. Similarly, if the additional algae and organics are removed before filtration, then chemical costs and sludge will increase. Higher temperatures likely will lead to poorer raw water quality but may increase treatment effectiveness. The net effect of climate change on treatment processes will need to be evaluated locally by each utility.

ADDRESSING CLIMATE CHANGE AND WATER RESOURCES MANAGEMENT

Climate change effects on water resources need to be managed through short- and long-term mitigation and adaptation strategies. These strategies are responses to either mitigate adverse effects or adapt to the "new normal" (changed water resources) conditions. Mitigation refers to actions that may be taken by a water utility to address the causes of potential climate change. For water utilities, mitigation actions may include reducing the carbon footprint of the treatment and distribution system by (1) reducing energy needs, (2) using more energy-efficient transportation, (3) implementing renewable energy sources, and (4) changing business and operational practices.

Adaptation refers to actions and responses that may be taken to address potential effects from climate change. Such responses to anticipated climate change effects may include a new basis for yield analysis of water sources, new water treatment and conveyance designs and standards, different source management and treatment operating rules, development of new infrastructure, enhanced stormwater management practices, increased demand management, innovative water allocation arrangements, and perhaps new water laws and regulations.

Water resources planning and management will require greater focus on risk management, integrated resources planning, and holistic or systems thinking as well as decision-support tools specific to climate change that can be included in integrated modeling systems to manage regional water resources. Such planning requires an understanding of relationships among physical factors, water resources, infrastructure, regulations, and operational practices. Modeling these relationships allows resource managers to evaluate and assess the potential effects of changes in physical factors and identify future utility practices and policies to respond to changing conditions. Figure 1-4 shows the relationship between climate change effects, effects on water resources, and management responses to address those effects.

Mitigation and Adaptation Techniques

Because the sustainability of our water resources will be influenced by the effects of climate change, it would be prudent for utilities to start thinking about appropriate mitigation and adaptation techniques to tackle upcoming challenges. Adaptation measures will typically be based on specific climate change conditions in the local region, which will require analyzing projected climate conditions to evaluate a specific utility's vulnerabilities. Mitigation measures, on the other hand, may be driven as much by laws and regulations as operational constraints.

Many of the mitigation and adaptation techniques presented in the remainder of the chapter are good practices that most utilities should consider adopting and are discussed in AWWA manuals of practice. However, climate risk and vulnerability should be understood and incorporated into these practices to improve a utility's resiliency to climate threats.

Long-Term Planning for Water Supply and Water Quality

While AWWA Manual M50, *Water Resources Planning* addresses long-term planning for water supply and water quality, climate change and the consideration of paleoclimate reconstructions are expected to strongly influence variables that in turn can create substantial risks to a water utility's direct operations that must be addressed. Climate change is expected to affect the quantity, quality, timing, and distribution of water available to utilities. Climate change–induced shifts within the hydrologic cycle are expected to pose substantial effects to both the reliability and quality of water supplies.

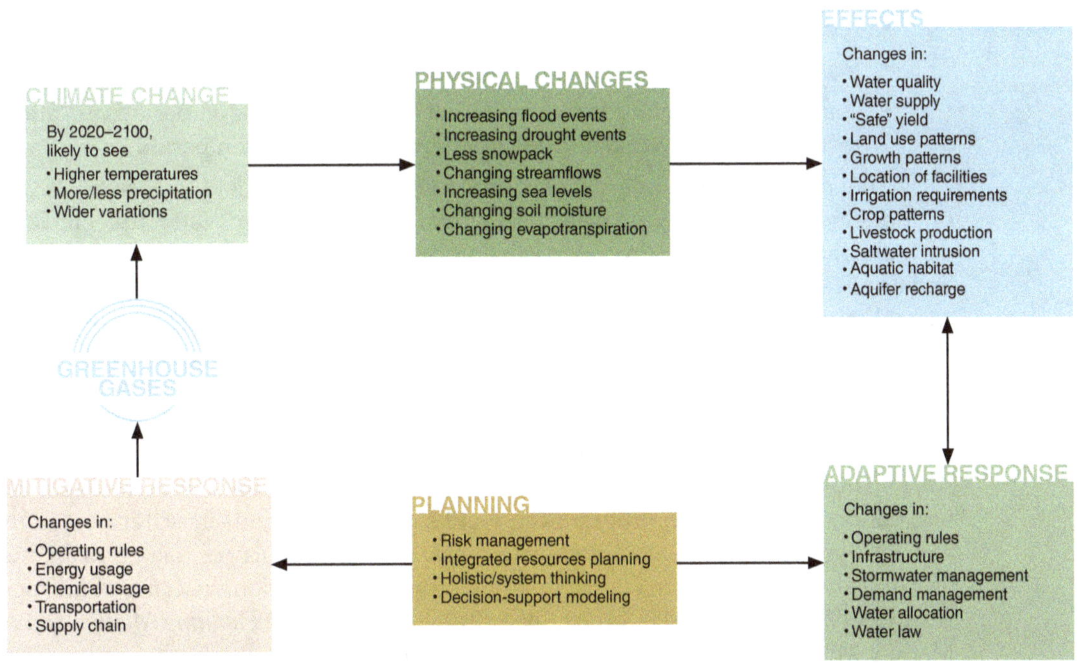

Figure 1-4 Relationship between climate change effects, effects on water resources, and management responses

However, the uncertainties associated with climate change need to be put in the context of multiple uncertainties and vulnerabilities that water utilities deal with on a regular basis, such as growth projections, changes in regulations, changing water consumption patterns, catastrophic events, and financing challenges.

From a water utility's perspective, almost all climate change–induced effects have the potential to severely alter the level of service provided to customers. Although it may not be possible to precisely predict the extent, location, timing, and rate of potential climate change effects, methods exist for developing a range of potential future hydrologic conditions that can help utilities develop flexible and adaptive management strategies. Specialized decision-support methods are also required to enable planners and managers to make long-term decisions given the significant uncertainty in climate projections. Further, long-term plans should be periodically updated based on our evolving understanding of climate change and changing service area conditions for utilities to continue to provide an adequate level of service capable of meeting projected demands well into the future. For more information on long-term planning, please refer to AWWA Manual M50, *Water Resources Planning*.

Water Demand Management Options

Many utilities have already realized the benefits of integrating water demand management into their operations. Demand management techniques such as leak detection and water conservation programs have proven successful for many utilities by postponing the need for infrastructure expansion through reductions in water demands by its customers. Demand management tools, approaches, and techniques can be adapted to a range of local water management conditions. These techniques typically focus on the dual priorities of improving water use efficiency and maximizing the effective reallocation of stored water. Examples of demand management options include

- reducing the quantity of water required to accomplish a specific task or service;
- managing and reducing water quantity system losses from source to tap; and
- shifting peak demand for water so that high water demand periods are attenuated, resulting in a more balanced water demand over a given period.

The unifying principle among the demand management measures is the goal of reducing and controlling water demands and improving water use efficiency. For agricultural zones, a shift to less water-intensive crops or more efficient water delivery and application practices may reduce local demand for water. For urban/suburban areas, controlling outdoor water use may be important to sustaining water resources. Reducing outdoor water use may be accomplished through incentive programs such as cash for grass programs, regulatory controls such as watering restrictions, or conservation rate structures. Long-term controls require an integrated planning approach in which, for example, subdivision controls limit lawn area, and certain irrigation technologies, such as soil moisture sensors, are required for all new development. For utilities, an aggressive loss prevention/leak detection program can yield significant savings in water wasted through distribution system losses and changes in basic operating practices related to maintenance and delivery of water from source to tap. For more information on demand management, refer to AWWA Manual M52, *Water Conservation Programs—A Planning Manual* and AWWA Manual M36, *Water Audits and Loss Control Programs*.

Operational and Management Controls to Revise Current Utility Operations

Several operational and management controls are available to assist water utilities in responding to predicted effects associated with climate change. Vulnerability analyses can be undertaken to identify the most urgent priorities in processes in which effects could be most detrimental to a utility's level of service. The changes in operational and management practices may include revision of reservoir operating procedure, drought management rule, or interagency agreements to tackle emergency situations. This may also require educating the customers, applying restrictions on water use type and quantities for large commercial and industrial users, or encouraging use of non-potable or reclaimed water for irrigation and toilet flushing. This will also require utilities to develop forward-thinking rate structures to encourage water conservation while recovering all utility costs, particularly if overall water usage goes down.

Integrated resource planning (IRP) can also be applied to a utility's current and future operations. The IRP approach has been used extensively in the water resource industry and often integrates environmental, socioeconomic, and engineering considerations into water supply planning. These considerations and their interactions are managed to provide maximum flexibility and resiliency within the water supply and distribution system. For more information on the IRP approach, see AWWA Manual M50, *Water Resources Planning*.

Reduce Carbon Footprint

The primary means for water utilities to reduce their contributions to climate change is to reduce GHG emissions from water utility operations. The Water Research Foundation and the California Urban Water Agencies (CUWA) published a study in 2009 (WRF 2009) that provides comprehensive guidance on standard protocols for quantifying direct and indirect GHG emissions and options to reduce those emissions. The following is drawn from that study.

> *There are many reasons that water utilities may begin to assess the emissions caused by their operations and to evaluate the projects or actions that may reduce or offset those emissions. These include, but are not limited to, existing and developing regulations requiring inventory and reporting of emissions, developing "carbon tax" and "cap-and-trade" regulation of GHG emissions, stakeholder initiatives demanding action, and environmental stewardship goals.*

Suggested procedures for development of an overall climate change strategy for a water utility begin with an estimation of GHG emissions inventory and footprint, as well as the potential effects of climate change on the utility (e.g., a change in source water supply). After the development of an inventory and footprint, potential internal and external projects for mitigation of effects can be identified and assessed. Potential internal project types may include energy efficiency efforts, addition of onsite renewable generation, and fleet management. The breadth of potential external project types is much greater and may include water conservation projects, terrestrial or geologic sequestration, energy efficiency, and many other project types that could be pursued through direct funding or via carbon brokers. After potential project types are identified, an overall reduction strategy can be developed and funding appropriated.

Energy Management

Effective energy management is essentially the process of maximizing efficiency while minimizing energy demand and reducing costs and GHG emissions. It applies to all aspects of business that consume energy including new construction, facility renovations, purchasing activities, operations, and maintenance. Energy efficiency may be the cheapest option and may provide a utility the best return on investment. Certain approaches such as using renewable energy through solar energy, have the potential to greatly reduce the environmental effect and GHG emissions associated with plant operations.

Other approaches target reducing the on-site consumption of energy through improvements to energy-intensive equipment and processes including pumps, motors, lighting, blowers, and aerators. Approaches to limit process energy requirements may include power monitoring, controls, chemical balancing, and other process improvements. As part of an overall energy management strategy, one must understand operational constraints while analyzing existing rate tariffs and comparing them against alternatives.

Although activities such as energy improvement projects or supply-side energy analyses are important aspects of an energy management program, the overall energy and carbon reduction potential is greatly increased when energy becomes a primary consideration in all aspects of operations. As an example, replacement motors or lights might be purchased on a regular basis. Periodically reviewing available options to identify more efficient alternatives may yield significant energy savings over time. Expansion projects and repair and replacement activities provide opportunities to greatly enhance energy performance. By developing a facility-wide energy management initiative and tracking and reporting results on a regular basis, the potential for achieving sustainable energy savings is greatly increased.

Regional Cooperation

Regional cooperation is important when evaluating climate change effects and developing adaptation strategies because the effects will cross multiple jurisdictional boundaries. Regional planning agencies and state water resource agencies are well-positioned to assist in efficiently evaluating climate change impacts to multiple agencies in a coordinated effort. For example, a regional planning agency could work with state or federal agencies to develop revised rainfall frequency analyses and drought frequency analyses that could then be used by multiple agencies, generating consistent planning assumptions across multiple agencies. Similarly, some response actions to buffer climate effects, such as increased raw water storage, may be best developed through a cooperative regional effort. This cooperative effort may be more successful than a localized agency effort in overcoming the challenges of siting and permitting new storage facilities. Adding regional emergency interconnections between water systems can also reduce risk of inadequate supplies during extreme droughts by balancing demands and supplies across a larger pool of sources and users.

Supplemental Water Supplies

For some utilities, management and operational modifications may not be sufficient to address effects to existing supplies, in which case additional supplies will be required to meet demands. In some cases supply augmentation will only be needed occasionally to supplement supplies during severe drought conditions and can be met with short-term solutions, such as purchasing water rights options from agricultural interests or other cooperative agreements. However, other utilities may need to pursue permanent supplies to meet their needs and to maintain the reliability of supply. In either case, significant investment in infrastructure is often required and may take many years to develop. Further, many localities face an array of political and financial constraints that may increase the time and effort required to implement any plan. Given the difficulty in implementing large water-supply projects and the level of uncertainty associated with climate change, utilities may find that increasing their water supply portfolio with several small projects will allow them to diversify supplies and hedge future uncertainties while costing less and being more politically feasible. Drawbacks may include significant increases in complexity of the system, which may require more sophisticated forecasting and planning tools to help inform decision-makers to optimize operations and meet a diverse range of objectives (e.g., cost, regulatory, public health, environmental).

Increased Awareness

Although water utilities routinely assess threats and vulnerabilities to their systems and source water supplies, it is critical to emphasize the need to include the probabilities and risks associated with water-related climate change effects and the possible range of adaptations available to respond to these changes. Multi-utility and multidisciplinary knowledge is being developed that can help streamline these efforts and provide useful tools for assessing risks and identifying approaches to address climate change effects. The need for increased awareness also extends outside the water industry to include the general public. Planning decisions and mitigation measures driven by effects posed by climate change will affect public policy and budgets. The inclusion of climate change considerations in community and critical water infrastructure planning will require a greater understanding of climate change, its effects, and the costs to adapt to those effects. Difficult climate-related decisions by utilities will be easier to make if there is sufficient political will driven by a well-informed community.

Other Supporting Management Measures

Other management measures supporting climate change adaptation include reducing, reusing, and recycling water and non–climate-dependent sources such as seawater desalination. Water demand reduction can be achieved through water use efficiency techniques applicable to both indoor and outdoor settings. Some of the indoor techniques include high-efficiency toilets and waterless urinals, water-efficient appliances (WaterSense labels on dishwashers, clothes washers), and leak repair. Some of the outdoor techniques include native and drought-resistant plant species selection (xeriscaping); use of artificial grass and efficient watering with drip systems, including generating daily and hourly weather data for use in water management practices; micro misters; and subsurface irrigation systems.

Rainwater capture refers to the collection of rainwater for outdoor irrigation and internal nonpotable uses through the use of cisterns or other storage techniques. Condensation collected from air conditioners and cooling towers could be considered for reuse as well. Rain gardens and green roofs can be established, and rainwater can be diverted or percolated via porous pavement and curb cuts. Gray water reuse involves collecting water that is discharged down interior drains such as sinks and showers, treating it with on-site systems, and reusing it for nonpotable indoor and outdoor uses, such as for toilets and irrigation. Gray water reuse is limited in some areas of the United States by regulations aimed at protecting against potential public health effects associated with reuse of untreated domestic water.

Water reclamation involves reusing treated municipal wastewater (i.e., reclaimed water). Reclaimed water can be used for outdoor irrigation, process cooling, and indoor nonpotable uses such as toilet flushing. In addition, advanced treated recycled water can be used to augment groundwater supplies and establish saltwater intrusion barriers. Reclaimed water is also being used in the United States and abroad to replenish drinking water aquifers and reservoirs (indirect potable reuse). A landmark project in Singapore, the NEWater project, returns treated municipal wastewater for direct use as drinking water (direct potable reuse). As fresh water supplies become increasingly scarce in many parts of the United States and the world, water reclamation should be considered a critical element of a sustainable water supply portfolio. For more information on water reclamation, see AWWA Manual M24, *Planning for the Distribution of Reclaimed Water*.

There are various challenges and threats posed by climate change and varied options for addressing the threats. The path forward for any utility or entity faced with climate change–related concerns is to develop an entity-specific evaluation of climate change impacts, system vulnerabilities, and available adaptive and mitigation measures to address the threats and vulnerabilities by means of a CAP. The remainder of the manual focuses on the information required to prepare a robust CAP.

REFERENCES

(AWWA) American Water Works Association. 2019. M24, *Planning for the Distribution of Reclaimed Water*. Denver: AWWA.

AWWA. 2017. M50, *Water Resources Planning*. Denver: AWWA.

AWWA. 2017. M52, *Water Conservation Programs—A Planning Manual*. Denver: AWWA.

AWWA. 2016. M36, *Water Audits and Loss Control Programs*. Denver: AWWA.

CH2M HILL. 2009. Confronting Climate Change: An Early Analysis of Water and Wastewater Adaptation Costs. Association of Metropolitan Water Agencies, 2009. Englewood, Colo: CH2M HILL.

Huxley, D.E., Bellamy, W.D., Sathyanarayan, P., and Ridens, M. 2009. Greenhouse Gas Emission Inventory and Management Strategy for Water Utilities. The Water Research Foundation (formerly AwwaRF) (WRF) and California Urban Water Agencies (CUWA). 2009. Englewood, Colo: CH2M HILL.

IPCC (Intergovernmental Panel on Climate Change). 2014. Climate Change 2015: Synthesis Report. Contribution of Working Groups I, II and III to the Fifth Assessment Report of the Intergovernmental Panel on Climate Change [Core Writing Team, R.K. Pachauri and L.A. Meyer, editors.]. Geneva, Switzerland: IPCC, 151 pp.

IPCC. 2007. Climate Change 2007: Synthesis Report. Contribution of Working Groups I, II, and III to the Fourth Assessment Report of the Intergovernmental Panel on Climate Change [Core Writing Team, R.K. Pachauri and A. Reisinger, editors.]. Geneva, Switzerland: IPCC, 104 pp.

Littell, J.S., Elsner, M.M., Whitely Binder, L.C., and Snover, A.K. 2009. The Washington Climate Change Impacts Assessment: Evaluating Washington's Future in a Changing Climate – Executive Summary. In: The Washington Climate Change Impacts Assessment: Evaluating Washington's Future in a Changing Climate. Seattle: Climate Impacts Group, University of Washington.

NASA. Global Climate Change: Vital Signs of the Planet. climate.nasa.gov/vital-signs/global-temperature/ (accessed Feb. 13, 2020).

Rice, J.L., Woodhouse, C.A., and Lukas, J. 2009. Science and Decision Making: Water Management and Tree Ring Data in the Western United States. *Journal of American Water Resource Association.* 45(5):1248-1259.

United State Department of Agriculture (USDA). Sustainable Watersheds, June 2008, www.fs.fed.us/kidsclimatechange/products/water-climate-brochure.pdf (accessed April 5, 2020).

Chapter 2

Purpose and Audience

The purpose of this manual is to help water and wastewater utilities plan for the short- and long-term impacts of climate change by guiding them through the development of a climate action plan (CAP). This manual can be used by water utilities, wastewater utilities, consultants, research institutions, academic institutions, and other organizations that may develop a CAP. A CAP may be a stand-alone document or developed as part of a master planning effort. This manual will help the user determine whether a CAP would be beneficial to a utility and identify the appropriate approach to develop it. Each chapter of this manual may be used as an independent resource. The manual also provides case studies that show examples of different approaches to developing a CAP as well as references for additional studies.

This manual presents a proactive voluntary approach to CAP development, which has typically been driven by local government agencies or cities. Additionally, the manual highlights adaptation rather than mitigation as there is already a significant body of work in the scientific and engineering community on the mitigation aspects of climate change.

This manual was developed through voluntary efforts by experts across the United States and around the world. The primary objective was to focus on North America, and the authors have sought to address climate change impacts, responses, and strategies throughout that continent. While the manual focuses on North America, information applicable to other countries is included as well.

DEFINITIONS OF KEY TERMS

Adapt, adaptation: (1) Adjustment in natural or human systems to a new or changing environment that exploits beneficial opportunities or moderates negative effects (NRC 2010a-d). (2) Implementation of measures intended to reduce the vulnerability of natural and/or human systems against actual or expected climate change effects. Various types of adaptation exist, e.g., anticipatory and reactive, private and public, and spontaneous and planned (IPCC 2009).

Adaptive capacity: The ability of a system to adjust to climate change (including climate variability and extremes) to moderate potential damages, to take advantage of opportunities, or to cope with the consequences (NRC 2010a-d).

Conservative estimate: One that provides for the full range in quantity or degree of an impact due to climate change in order to assess the full range of potential consequences, damages, or opportunities.

Mitigation: (1) An intervention to reduce the causes of changes in climate, such as through reducing emissions of greenhouse gases (GHGs) to the atmosphere (NRC 2010a-d). (2) Technological or operational change and/or substitution that reduces the forcing of the climate system (i.e., reduces resource inputs and GHG emissions per unit of output). Although several social, economic, and technological policies would produce an emission reduction, with respect to climate change, mitigation means the *implementation* of technological or operational change and/or substitution to reduce GHG emissions and enhance sinks (IPCC 2007).

Resilience: A capability to anticipate, prepare for, respond to, and recover from significant multi-hazard threats with minimum damage to social well-being, the economy, and the environment (NRC 2010a-d).

Risk: A combination of the magnitude of the potential consequence(s) of climate change impact(s) and the likelihood that the consequence(s) will occur (NRC 2010a-d).

Stationarity: The idea that natural systems fluctuate within an unchanging envelope of variability (NRC 2010a-d).

Uncertainty: Ranges from confidence just short of certain to informed guesses or speculation. Some categories of uncertainty are quantifiable, while other categories cannot be expressed readily in terms of probabilities. Uncertainties arise from such factors as linguistic imprecision, statistical variation, measurement error, variability, approximation, subjective judgment, and disagreement (Schneider et al 2002).

Vulnerability: The degree to which a system is susceptible to, or unable to cope with, adverse effects of climate change, including climate variability and extremes. Vulnerability is a function of the character, magnitude, and rate of climate variation to which a system is exposed, its sensitivity, and its adaptive capacity (NRC 2010a-d).

REFERENCES

IPCC (Intergovernmental Panel on Climate Change). 2007. *Climate Change 2007: Synthesis Report*. Contribution of Working Groups I, II and III to the Fourth Assessment Report of the Intergovernmental Panel on Climate Change [Core Writing Team, Pachauri, R.K and Reisinger, A., editors]. Geneva, Switzerland: IPCC, 104 pp.

National Research Council. 2010. "Advancing the Science of Climate Change." Washington, DC: The National Academies Press.

Schneider, S.H., Rosencranz, A., Niles J.O., editors. 2002. *Climate Change Policy: A Survey.*, Washington D.C.: Island Press. https://stephenschneider.stanford.edu/Publications/PDF_Papers/Ch02ClimatePolicy.pdf (accessed April 5, 2020).

Chapter 3

Science of Climate Change

Climate change is such a broad topic that it is easy to get lost in the available literature if one does not narrow the focus. This chapter reviews the literature on the science of climate change and summarizes the information relevant to hydrological impacts.

CLIMATE CHANGE AND CLIMATE VARIABILITY

Climate change is not just "changing weather." *Weather* refers to atmospheric conditions at a specific location at a given point in time (NWS Glossary, n.d.). In contrast, climate is defined as regional weather conditions averaged across a period of years (NWS Climate Prediction Center, n.d.). Generally speaking, climate is the statistical distribution of weather over the long term, typically interpreted as at least 30 years. Climate change refers to a change in the statistical distribution—the state of the climate—over decades.

Over millennia, the climate has varied due to natural forcings, such as volcanic eruptions and changes in solar irradiance. Patterns of internal climate variability occur due to the exchange of heat, water, energy, and gases between the atmosphere, ocean, cryosphere, and land surface. For example, the El Niño–Southern Oscillation (ENSO) is a multiyear pattern created by fluctuating sea surface temperature anomalies in the Pacific Ocean. ENSO directly influences storm tracks, temperature, and other weather events across the globe (NOAA's El Niño Portal 2015). Other modes of variability include high- and low-frequency phenomena like the Pacific Decadal Oscillation (PDO), the North Pacific Oscillation (NPO), the North Pacific Gyre Oscillation (NPGO), the North Atlantic Oscillation (NAO), the Madden-Julian Oscillation (MJO), Atlantic Multidecadal Oscillation (AMO), and the Indian Ocean Dipole (IOD).

At regional scales and short time periods, internal climate variability can dominate climate (Melilo et al. 2014). For example, La Niña (the cool phase of ENSO) can still occur even though global average temperatures are increasing due to climate change. Such climate variability will continue to modify the pace of climate change in regions like the Pacific Northwest of the United States for the next several decades (Abatzoglou et al. 2014). These processes will continue to drive seasonal and interannual changes in temperature and precipitation and can magnify or moderate the effects of long-term climate trends in a given region.

Although these natural modes of variability dominate in the short term, external forcings on the climate system in the form of anthropogenic greenhouse gas (GHG) emissions are the primary cause of current global warming. Atmospheric concentrations of GHGs such as carbon dioxide, methane, and nitrous oxide have increased since the pre-industrial era, driven largely by economic and population growth (IPCC 2000). The heat-trapping effects of these GHGs, together with the impact of other human activities (like deforestation) have been detected throughout the climate system and are extremely likely to have been the dominant cause of the observed warming since the mid-20th century (IPCC 2000). Figure 3-1 illustrates the change in global temperature records over the last century. Though temperatures fluctuate year to year, it is clear that the average global temperature has increased over the past century.

GLOBAL CLIMATE MODELS AND GREENHOUSE GAS EMISSIONS

To project how GHG emissions and other anthropogenic forcings will affect the climate system, climate scientists use atmosphere–ocean general circulation models (AOGCMs) or general circulation models (GCMs). The AOGCMs/GCMs are three-dimensional, computer-based mathematical representations of atmospheric, oceanic, and land surface

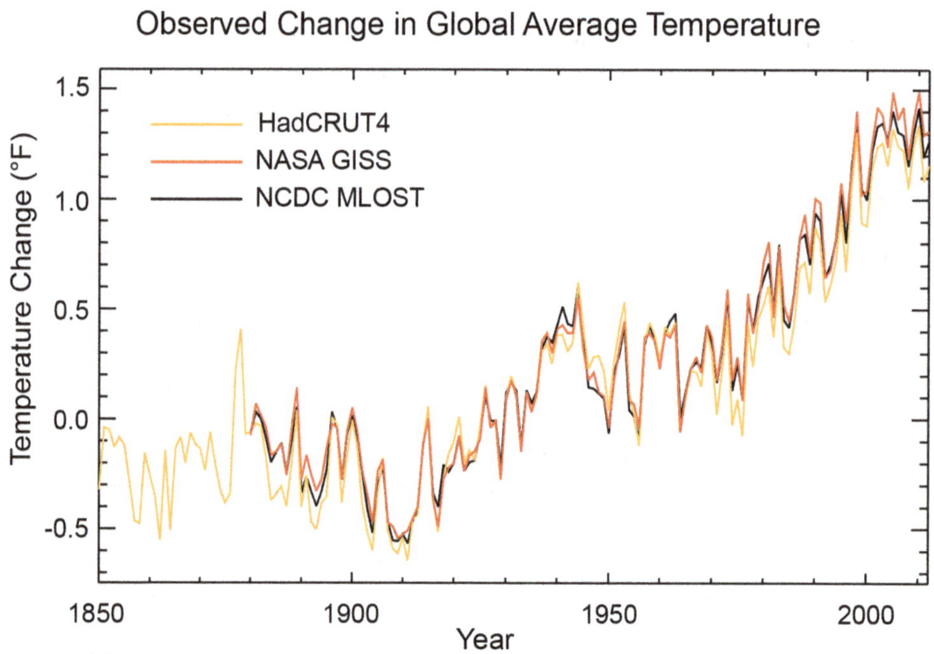

Source: Peter Thorne, Nansen Environmental and Remote Sensing Center; adapted by Third National Climate Assessment, 2014
HadCRUT4—Global Temperature Dataset developed by Climate Research Unit and Hadley Center (United Kingdom), NASA GISS—National Aeronautics and Space Administration Goddard Institute for Space Studies, NCDC MLOST—National Oceanic and Atmospheric Administration Merged Land-Ocean Surface Temperature Analysis

Figure 3-1 Observed change in global average temperatures from three global surface temperature records

processes based on the fundamental laws of physics and mathematical discretizations that resolve atmospheric and oceanic processes (IPCC 2007). GCMs typically divide the globe into grid cells of around 100–200 km^2 and are able to simulate certain physical parameters well at each grid cell, but other processes that occur at finer scales often need to be parameterized or estimated (e.g., clouds, aerosols, thunderstorms). Over recent decades, GCMs have become better able to simulate Earth's observed climate, including temperature, precipitation, radiation, wind, ocean temperatures, ocean currents, and sea ice cover (IPCC 2007).

To project future conditions, GCMs are "forced" with different boundary conditions; in particular, GHG emissions scenarios. Emissions scenarios are developed by the Intergovernmental Panel on Climate Change (IPCC). Currently, two different types of future emissions scenarios are widely used in the global climate modeling community.

The first category of scenarios comes from the Special Report on Emissions Scenarios (SRES; IPCC 2000). These scenarios were used in the IPCC Third Assessment Report (TAR) published in 2001, and in the IPCC Fourth Assessment Report (AR4) published in 2007. The SRES represent different future global economic, technological, or population storylines (IPCC 2000).

More recently, Representative Concentration Pathways (RCPs) were used in the 2014 IPCC Fifth Assessment Report to quantify the amount of future solar radiative forcing. The four RCPs represent a broad range of climate outcomes: stringent GHG mitigation scenario (RCP2.6) in which emissions decrease before mid-century and the change in global temperature above pre-industrial levels is kept below 2°C; two intermediate GHG scenarios (RCP4.5 and RCP6.0) where emissions stabilize and then decrease by mid-century or increase through mid-century but then decrease by late-century, respectively; and one scenario with very high GHG emissions that increase through the end of the century (RCP8.5). RCP 8.5 is considered similar to a continuation of the current path of global emissions increases in the absence of significant reductions in global GHGs (Melillo et al. 2014). These pathways are shown in Figure 3-2.

The SRES were applied to "force" the generation of GCMs from the Coupled Model Intercomparison Project 3 (CMIP3) that were used for IPCC AR4, and the RCPs were used more recently in the Coupled Model Intercomparison Project 5 (CMIP5) model runs for the IPCC Fifth Assessment Report. CMIP is a program of standardized global model experiments and comparisons (WCRP 2015) that include experiments from several "families" of climate models—models that were developed by the same scientific institution or modeling research center. The CMIP models include an ensemble of more than forty different models, several of which are from the same family.

GLOBAL CLIMATE MODEL PROJECTIONS AND ASSUMPTIONS

Though GCMs provide credible quantitative estimates of future climate change, particularly at continental and global scales, there is still uncertainty within the models. One source of uncertainty is the quantity of future emissions—generally the choice of emissions scenarios becomes more important toward the end of this century (Northrop and Chandler 2014) when different emissions choices lead to diverging warming projections. The choice of emissions scenario can significantly impact the model's future temperature and precipitation projections.

Another key assumption is that a model's ability to simulate observed climate indicates the model's ability to project future climate. This cannot be proven, especially because the reliability of future projections does not account for positive or negative feedbacks in the climate system that could change future outcomes. The climate is a chaotic system: small changes in physical processes in one part of the climate system can have significant

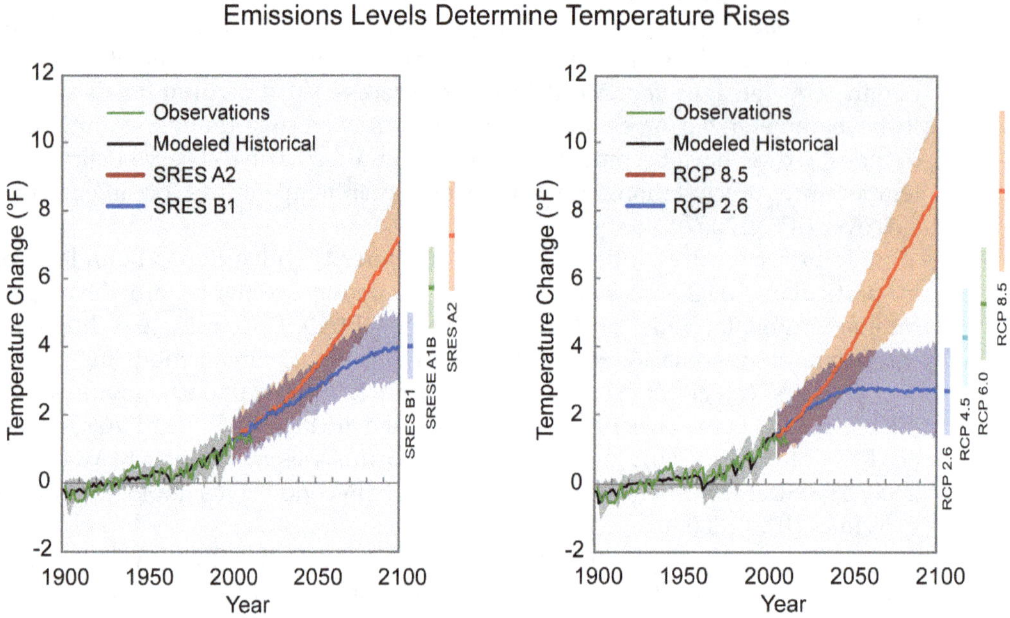

Source: Melillo et al. 2014

RCP—Representative Concentration Pathways, SRES—Special Report on Emissions Scenarios

Figure 3-2 Emissions levels using SRES or RCP scenarios determine the amount of global temperature change

Notes:

Observations: temperature observations.

Modeled historical: Temperature change modeled using climate models for a historical period of 1900 to early 2000s. Values in this plot represent an average of all climate models.

RCP 8.5: Temperature change modeled using climate models for a future period from 2020–2100. A worst-case greenhouse gas emission scenario represented by RCP 8.5 is used to represent future socioeconomic conditions.

RCP 2.6: Temperature change modeled using climate models for a future period from 2020–2100. A best-case greenhouse gas emission scenario represented by RCP 8.5 is used to represent future socioeconomic conditions.

impacts on another part of the climate system. Because there is no way to experiment with the Earth's climate in a laboratory, scientists do not know for sure if a model's performance in replicating historical climate is a perfect indication of its ability to project future climate.

Furthermore, climate models are not probabilistic—they cannot predict the future definitely, and no single model provides a more likely prediction than another. Instead, models project a range of plausible different futures. Using a different model will yield different results. Because individual models have different strengths and weaknesses, it is best to use a suite of climate models to bracket a broader range of future projections. Scientists have also evaluated model performance in certain regions of the United States based on how accurately they reproduced temporal and spatial statistics of 20th century observed climate, and then ranked models relative to one another according to the set of performance metrics considered (Rupp et al. 2013).

Despite these uncertainties and assumptions, GCMs still offer the most scientifically robust approach to projecting future climate. Additional modeling efforts localize large-scale GCMs to yield projections that are more relevant to water utility planning. See Chapter 4 of this manual for more information on modeling and downscaling.

The following online resources provide more information on the scientific basis of climate change and GCMs:

- Climate Change 2013: The Physical Science Basis. Fifth Assessment Report Working Group 1 (IPCC 2013)

- Climate Change Evidence & Causes: An Overview From the Royal Society and the US National Academy of Sciences (2014)
- Climate Models: An Assessment of Strengths and Limitations, US Climate Change Science Program, Synthesis and Assessment Product 3.1 (2008)
- COMET MetED: Introduction to Climate Models
- US Global Change Research Program Third National Climate Assessment (2014)

CLIMATE CHANGE AND THE HYDROLOGIC CYCLE

How does a warming climate affect water utilities? Climate is a key driver of the water cycle and influences the timing, intensity, seasonality, and spatial distribution of water resources. Figure 3-3 illustrates how surface flow, runoff, snowmelt, infiltration, and groundwater flow are affected by climatic processes such as evaporation, precipitation, and condensation. A warmer atmosphere can hold more moisture, so water vapor acts as a GHG in the atmosphere, creating a positive feedback loop where more water vapor leads to more warming (The Royal Society and National Academy of Sciences 2014), compounding the warming effect of adding more GHGs to the atmosphere.

Warming is already leading to an intensification of the water cycle across North America. Hydrologic impacts from climate change include an increase in the frequency and intensity of storms and heavy rain events; changes in seasonal precipitation patterns; increased likelihood of flood events; reduced snowfall and snowpack; earlier snowmelt runoff; sea level rise; saltwater intrusion of coastal aquifers; decreased groundwater recharge and availability; decreased soil moisture; intensification of droughts; and deteriorating water quality due to increases in water temperature, nutrients, sediment, and pollutants (Melillo et al. 2014). These shifts in the water cycle have consequences for the reliability of water supplies and water quality for drinking water systems and the management of stormwater and wastewater systems.

Many of the impacts noted previously have already been observed and are expected to become more pronounced as climate warming continues. However, impacts will differ by region and type of water system. Local changes in climate can be very different from global projections. Critically, changes in climate and changes in hydrology are not always linearly connected—even small changes in average climate can produce large changes in extremes, according to paleohydrologic studies (Milly et al. 2008).

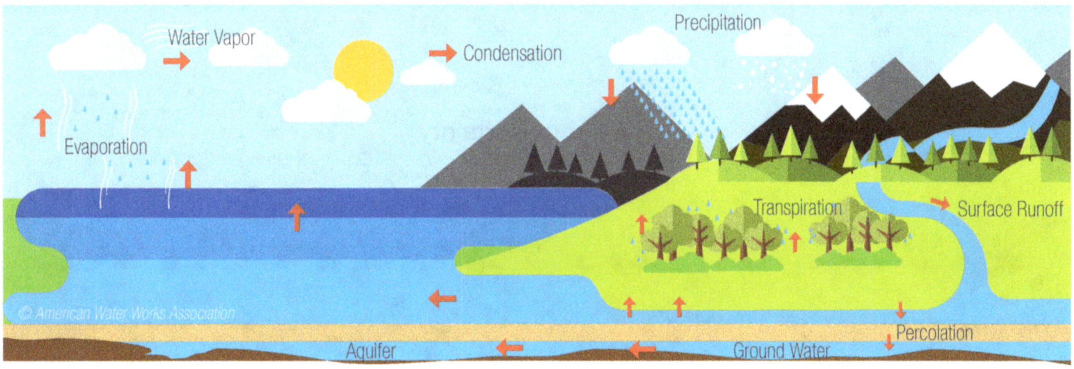

Source: DrinkTap.org

Figure 3-3 The water cycle

From Static to Dynamic Systems

In his book *Living With Water Scarcity*, David Zetland states that "climate change will increase the rate of assumption failure" (2014). This captures one of the main challenges that climate change presents to water utilities: how to plan within the context of a more uncertain future climate, where past is no longer prologue.

Water utilities have traditionally assumed some level of stationarity in short and long-term planning. Historical hydrology data and time series are considered sufficient representations of what to expect in the future. This approach assumes the average, extremes, and variability of climate and hydrology will not significantly differ in the future and is illustrated by a "cylinder of certainty" as shown in Figure 3-4.

But the science of climate change suggests that the assumption of a static climate and hydrology may be limited. Climate change is shifting conditions beyond the historical variability that water utilities have observed. Of concern to water managers is that climate change "threatens to make this variability greater, shifting and intensifying the extremes, and introducing greater uncertainty in the quantity and quality of supply over the long term" (Grayman et al. 2012).

The bell-curve distribution of historical "normal" climate in Figure 3-5 illustrates how the distribution of weather or climate is centered around average events which are the most likely to occur, but the climate system also infrequently experiences extreme events at each tail of the distribution (e.g., extreme storms, floods, hurricanes, or droughts). Many water utilities have traditionally planned within the historical range; however, climate change is shifting the entire distribution curve (Figure 3-5). Historical extremes will become the new "normal" conditions, and a set of new extremes will emerge. This does not indicate a shift from one static state to another, but rather a continuous change over time as the climate continues to shift and increase in variability.

These shifts in the distribution of both weather and climate will create a nonlinear climate and hydrologic system where small changes in the main drivers of the water balance—temperature, evapotranspiration, and precipitation—could lead to cascading impacts on peak runoff, snowpack, stream temperature, and other hydrologic conditions. Water utilities may no longer be able to rely on observed or historical weather to design and plan infrastructure and water supplies. Instead, utilities will need to think more dynamically about their systems in order to incorporate future climate uncertainty into planning.

Therefore, while climate modeling helps identify a spectrum of future changes, there is no way to know how accurate climate model projections will be. Water utilities may consider a range of plausible climate futures, depicted as the "cone of uncertainty" in Figure 3-6.

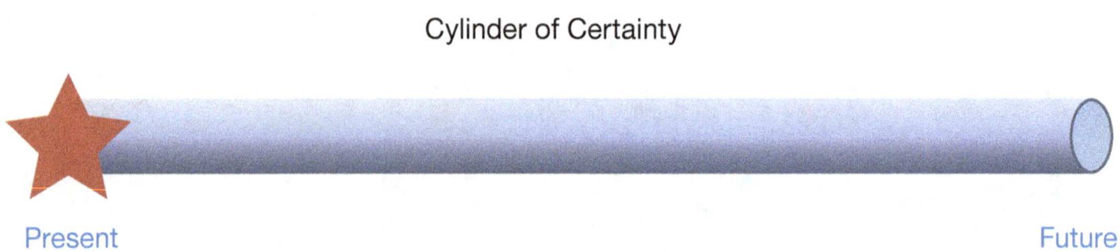

Source: Adapted from L. Kaatz, Denver Water, 2015

Figure 3-4 The "cylinder of certainty" illustrates how future conditions will be the same as present or past conditions

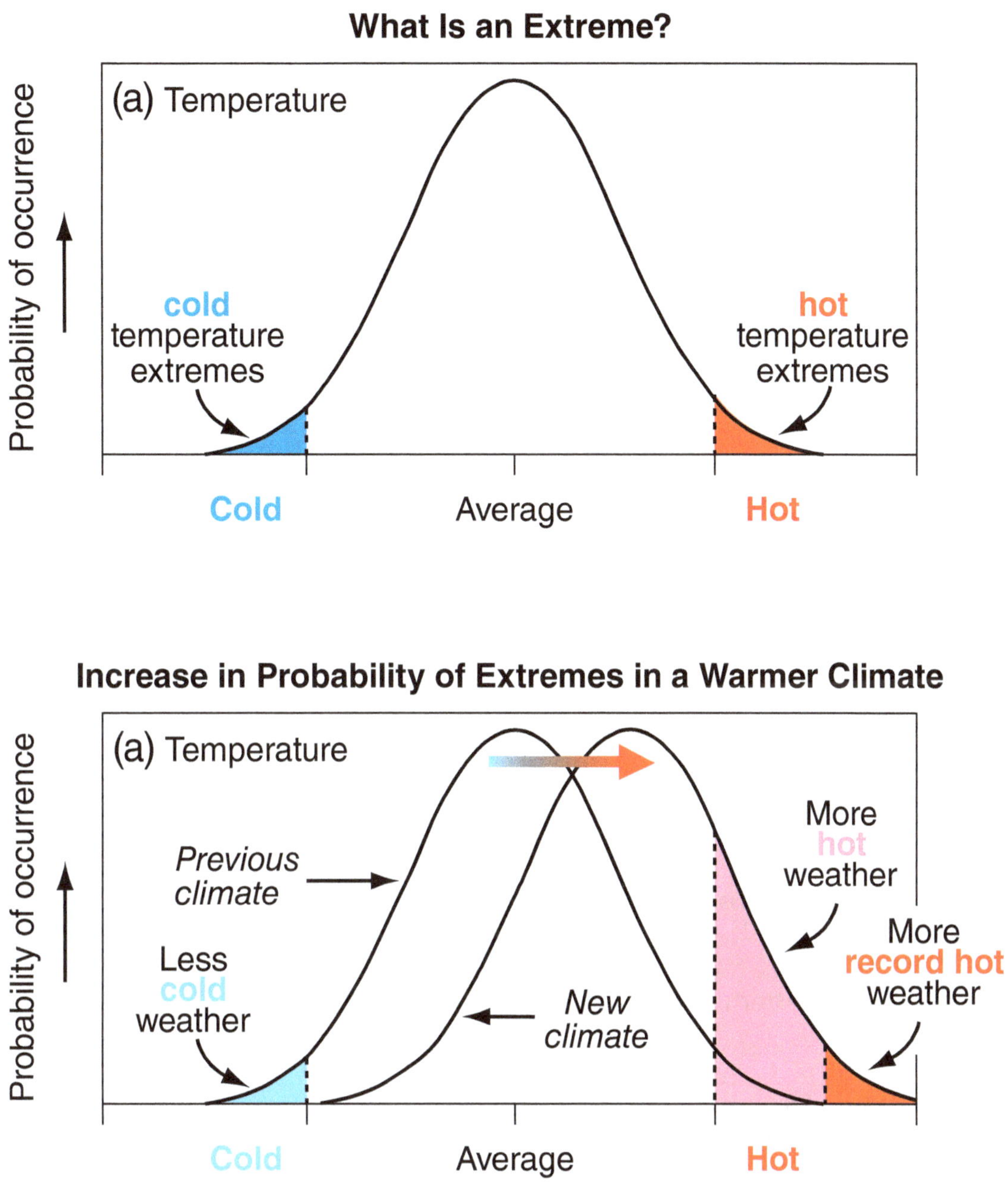

Source: IPCC 2001, Figure 2.32 of Climate Change 2001: The Scientific Basis. Contribution of Working Group I to the Third Assessment Report of the Intergovernmental Panel on Climate Change

Figure 3-5 "Normal" or historical weather distribution (top); how climate change is expected to shift the distribution of weather and climate (bottom)

While Chapter 4 provides more information about changing engineering practices in light of increased climate variability, water planners will find it helpful to first understand projected local climate change impacts on the vulnerability of specific water systems. The next two sections summarize key regional impact information and considerations for assessing the vulnerability of a specific system.

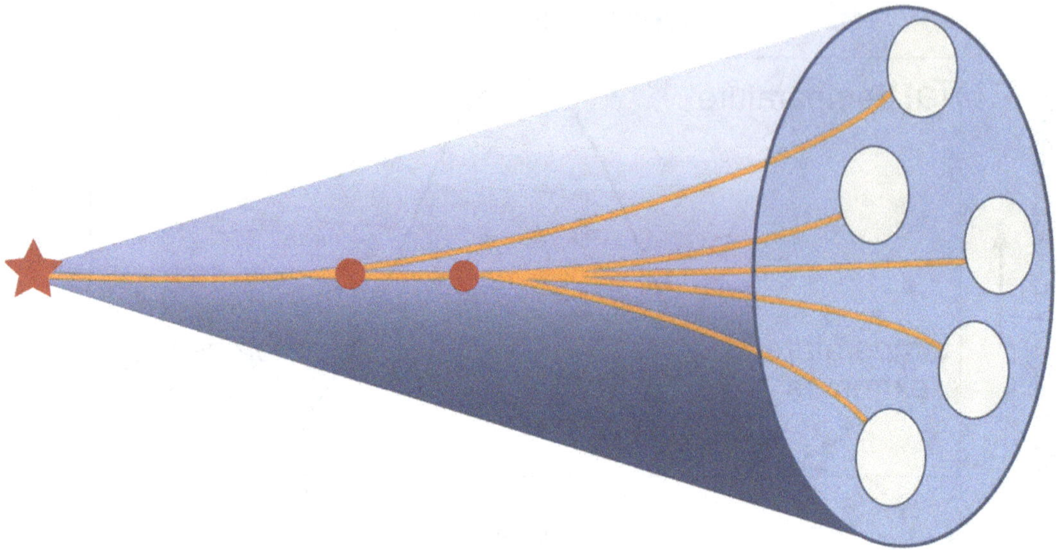

Source: Stratus Consulting & Denver Water 2015; adapted from L. Kaatz, Denver Water, 2015

Figure 3-6 The "cone of uncertainty" illustrates the concept of planning for multiple futures, in contrast to the "cylinder of certainty"

REGIONAL HYDROLOGIC IMPACTS OF CLIMATE CHANGE

This section highlights how climate change will impact hydrologic systems across different regions of North America. Specific summaries are provided for several regions of the United States, and broader summaries for Canada and Mexico. These summaries are not comprehensive, but they describe the general trends that are anticipated to affect water resources across most regions of North America. The resources listed at the end of this section provide additional information on regional climate impacts.

Southwestern United States

The southwestern region of the United States comprises the states of Arizona, California, Colorado, Nevada, New Mexico, and Utah. It is the hottest and driest region of the United States, where the availability of water has defined the region's landscapes, economy, and history of human settlement. The Southwest experienced its warmest decade from 2001 to 2010, with fewer cold air outbreaks, more heat waves, and temperatures almost 2°F higher than historical averages (Hoerling et al. 2013). Regional annual average temperatures are projected to rise due to climate change, with the greatest increases in the summer and fall months.

In contrast, precipitation trends vary considerably across the Southwest, with some areas experiencing decreases and others experiencing increases. Projections of precipitation changes are less certain than those of temperature (Kunkel et al. 2013a, Cayan et al. 2013). But seasonal changes, such as reduced winter and spring precipitation, are consistently projected for the southern areas of the Southwest by 2100. In the northern areas of the region, seasonal precipitation changes are smaller than natural variations (Kunkel et al. 2013b). Flooding may intensify in Southwest regions, even in areas where total precipitation is projected to decline. An increase in winter flood hazard risk in rivers is projected due to increases in flows of atmospheric moisture into California's coastal ranges and the Sierra Nevada (Dettinger 2011, Dettinger et al. 2011).

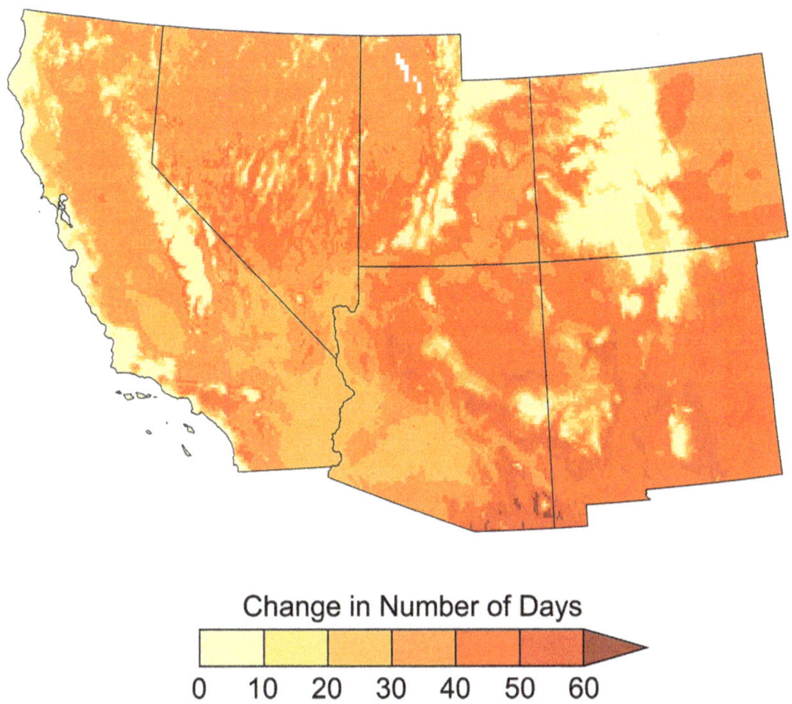

Source: US Global Change Research Program 2019, as adapted from Voss et al. 2017.

Figure 3-7 US drought monitor map of California for April 2019

Variability and uncertainty are the dominant characteristics of Southwest water resources. Geographic variation and unpredictable precipitation make it challenging to manage water needs. For example, California's water resources vary dramatically across the state because of extreme differences in precipitation (Dettinger 2011). Figure 3-7 shows that more than two-thirds of California is under the threat of "extreme" or "exceptional" drought due to a combination of lack of rainfall, lack of snowpack, and high temperatures. Climate change is expected to cause droughts and water supply challenges in California and other parts of the Southwest. Snow is a main source of water supply, but warmer temperatures are already causing more precipitation to fall as rain rather than snow (CDM 2011). This is leading to earlier spring snowmelt and runoff and decreasing surface water supply reliability.

The Southwest is historically prone to drought, and paleoclimate records show severe mega-droughts at least 50 years long (Cook et al. 2010; Meko et al. 2007; Routson et al. 2011). However, future droughts are projected to be substantially hotter. For major river basins, such as the Colorado River Basin and other parts of the region, drought is projected to become more frequent, intense, and longer lasting than in the historical record (Cayan et al. 2013), essentially intensifying with climate change. Climate change is also expected to affect water demand, groundwater withdrawals, and aquifer recharge, reducing groundwater availability in some areas of the region.

Additionally, regional Pacific Ocean sea level rise, extreme high tides, and storm surges are projected to affect California's coastal cities and infrastructure. Flooding and erosion in coastal areas is already occurring even at existing sea levels. Future sea level rise could increase the salinity of coastal aquifers in the state.

Southeastern United States and Caribbean

The southeastern United States and Caribbean region includes the states of Arkansas, Louisiana, Kentucky, West Virginia, Virginia, North Carolina, South Carolina, Kentucky, Tennessee, Mississippi, Alabama, Georgia, and Florida, as well as Puerto Rico and the US Virgin Islands. The region encompasses the Appalachian Mountains and the coastal plains, so the geographic distribution of climate impacts and vulnerabilities depends on location (Melillo et al. 2014). However, there are several common impacts that will affect large parts of the region. Future temperature rise is expected to cause significant increases in the number of hot days (95°F or above) and decreases in freezing events. Higher temperatures are also expected to increase harmful blooms of algae and several disease-causing agents in inland and coastal waters. The region is exceptionally vulnerable to sea level rise, extreme heat events, hurricanes, and decreased water availability due to projected climate change (Melillo et al. 2014). The number of Category 4 and 5 hurricanes in the North Atlantic and the amount of rain falling in very heavy precipitation events have increased over recent decades, and further increases are projected (Melillo et al. 2014).

The region, home to more than 80 million people, includes some of the fastest-growing metropolitan areas in the United States. Cities and water supplies along the Atlantic seaboard and Gulf Coasts are vulnerable to the impacts of regional sea level rise and storm surge, as shown in Figure 3-8. New Orleans—with roughly half of its population below sea level—and other major cities such as Miami, Tampa, Charleston, and Virginia Beach are among those at risk. Furthermore, as a result of current sea level rise, the coastline of Puerto Rico around Rincòn is being eroded at a rate of 3.3 feet per year. Puerto Rico has one of the highest population densities in the world, with 56 percent of the population living in coastal municipalities. Sea level rise and storm surge can have impacts far beyond the area directly affected.

Freshwater supplies from rivers, streams, and groundwater sources near the coast, especially porous aquifers, are also at risk of accelerated saltwater intrusion due to higher sea levels. Some cities have had to abandon their drinking water wells due to the effects of saltwater (Melillo et al. 2014).

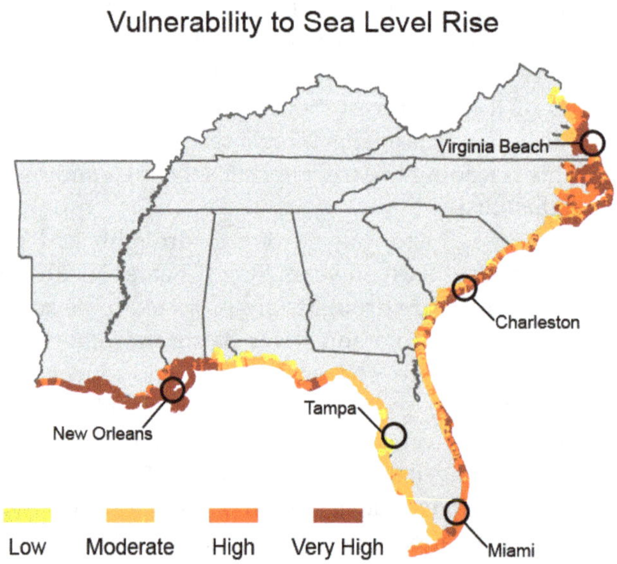

Source: US Global Change Research Program 2014

Figure 3-8 Vulnerability to sea level rise in the southeastern United States

In the future, several stressors are expected to reduce the availability of water in the Southeast. These include continued urban development, irrigated agriculture, and an increase in evaporative losses due to higher temperatures. Many small Caribbean islands are also likely to face severe water stress (Melillo et al. 2014).

The Midwest and Great Lakes

The Midwest and Great Lakes states of Minnesota, Wisconsin, Michigan, Ohio, Indiana, Illinois, Iowa, and Missouri are vulnerable to climate change. Direct effects will include increased heat, flooding, drought, and late spring freezes. Water quality impacts are a major concern in the region, in the form of increased watershed pollutants and toxic algal blooms due to higher temperatures and extreme weather events. Many cities have aging infrastructure and are particularly vulnerable to flooding. Heavy downpours have increased in the region, resulting in the discharge of untreated sewage into combined sewage–overflow systems in a number of Midwestern cities (Melillo et al. 2014). The region has previously experienced waterborne disease outbreaks due to drinking water source contamination from heavy precipitation and failing infrastructure (Patz et al. 2008). Extreme rainfall events and flooding due to climate change could exacerbate water quality challenges, cause erosion, and damage infrastructure. In particular, climate change risks to large water supplies like the Great Lakes, a source of drinking water for 40 million people (Alliance for the Great Lakes 2015), could manifest in harmful blooms of algae, declining beach health, and more invasive species.

The Great Plains

The Great Plains states—Wyoming, Texas, Oklahoma, Nebraska, Kansas, North Dakota, South Dakota, and eastern Montana—experience a diverse climate due to the region's large north–south spread and elevational gradient extending from sea level to the Continental Divide. Climate change impacts will therefore differ across the region.

Historically, the region has observed maximum temperatures of 95°F to 100°F over seven days or more on average. Continued warming means that these high temperatures are expected to occur much more frequently (Melillo et al. 2014). Seasonal winter and spring precipitation is projected to increase in the northern states. The central states are expected to experience small changes in summer and fall precipitation, with Texas and Oklahoma projected to suffer longer dry spells by mid-century. The frequency of heavy precipitation in the northern states is projected to increase by mid-century (Melillo et al. 2014).

Climate change will stress water resources in the region due to higher temperatures, more dry days and heat waves, increased drought frequency and intensity, and increased evaporation rates. Anticipated sea level rise along the Texas Gulf Coast will increase the risk of damage from storm surges in an area already prone to hurricanes and tropical storms (Melillo et al. 2014).

Northeastern United States

In the past several decades, the Northeastern states of Maine, New Hampshire, Vermont, Massachusetts, Rhode Island, Connecticut, New York, New Jersey, Pennsylvania, and Maryland have experienced higher temperatures and more extreme weather events. These changes are expected to escalate by the end of the century. Climate projections indicate that Northeast temperatures will rise by 2.5°F (1.4°C) to 4°F (2.2°C) in winter and by 1.5°F (0.8°C) to 3.5°F (1.9°C) in summer (Frumhoff et al. 2007). The number of days that exceed 90°F and the amount of precipitation are expected to continue to increase in the region (Frumhoff et al. 2007).

Unlike other regions of the United States where no clear change in annual precipitation is expected, the annual average precipitation in the Northeast is projected to increase between 7 and 14 percent by 2100. An 8 to 9 percent increase in precipitation intensity is projected by mid-century, climbing to 10 to 15 percent by the end of the century (Frumhoff et al. 2007). Heavy precipitation, defined as more than 2 inches of rain falling within 48 hours, is also expected to increase in both intensity and frequency, with the number of heavy precipitation events increasing by 8 percent by mid-century and 12 to 13 percent by the end of the century (Frumhoff et al. 2007). Precipitation changes affect the replenishment of groundwater supplies when soil becomes oversaturated, causing heavy rainfall to run off as surface flow rather than infiltrating into the soil. These conditions can cause outbreaks of waterborne diseases, as high soil saturation conditions increase the transportation of microbial organisms (Currerio et al. 2001). Coupled with projected sea level rise, the surge in precipitation frequency and intensity is anticipated to increase the risk of severe flooding events in this region. The US Environmental Protection Agency (USEPA) estimates that by the end of this century, New York City may experience a 100-year flood every 10 to 22 years (USEPA 2016).

As in other regions, rising temperatures have skewed the snow-to-rainfall ratio, with the ratio decreasing by as much as 30 percent in some Northeastern stations. Snowpack depth has declined, with some stations reporting a decrease of up to 16 percent in March/April snowpack (Hayhoe et al. 2007).

Northwestern United States

The Northwest states of Oregon, Washington, Idaho, and western Montana span a diverse range of geographic conditions and ecotones, including plateaus, high deserts, and coastal old-growth forests. The Cascade Range creates a strong rain shadow effect that divides the region into wetter and drier areas, and the Pacific Ocean influences regional climate variability through teleconnections like ENSO and the PDO. In the near term, this variability will continue to significantly influence Northwest climate.

However, the region has seen an increase in temperature since the late nineteenth century, despite seasonal and inter-annual variability. Future climate model temperature projections indicate that Northwest temperatures will rise year-round, with hotter summers and more incidences of extreme heat (Dalton et al. 2013). To date, annual mean precipitation has exhibited considerable variability and there has been no clear trend in extreme precipitation. Climate models do not agree on a change in amount of annual precipitation for the region. However, many models project an increase in extreme daily precipitation events over one inch, as well as a change in the seasonality of precipitation, with drier summers and wetter spring, fall, and winter seasons (Dalton et al. 2013).

Much of the low-elevation maritime snowpack in the region hovers around melting point 1°F (0°C) for most of the winter. Multiple studies show an observed decrease in mountain snowpack and snow water equivalent, as illustrated in Figure 3-9, and a shift toward earlier snowmelt and streamflow center of timing (Melillo et al. 2014, Hidalgo et al. 2009, Mote 2006, Nolin and Daly 2006). Even annual streamflows in the region show significant declines in the lowest quartile of flows because warming temperatures have led to less snow in snow-dominated watersheds (Luce and Holden 2009).

Warmer winter temperatures will cause more precipitation to fall as rain, leading to declining snowpacks, earlier spring snowmelt and peak runoff in snow-fed systems, and lower summer streamflows. The hydrologic regimes of watersheds during winter months are also likely to change: many mixed rain–snow (or transitional) watersheds are expected to shift to rain-dominant watersheds by the end of the century, with snow-dominant watersheds shifting to mixed rain–snow. Changing hydrographs will likely cause warmer stream temperatures that affect endangered species like salmon, decreased summer

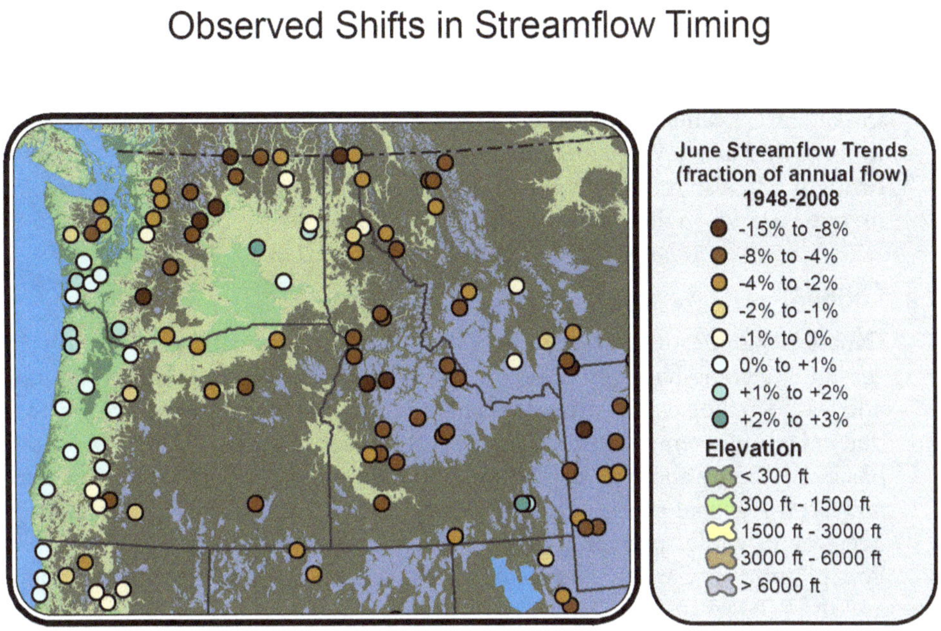

Source: US Global Change Research Program 2014, as adapted from Fritze et al. 2011

Figure 3-9 Reduced June flows in many Northwest snow-fed rivers is a signature of warming in basins that have a significant snowmelt contribution

water availability for instream uses and agriculture, and consequences for hydropower generation (Dalton et al. 2013, Melillo et al. 2014). In addition, an increase in extreme storms and the changing rain–snow ratio in snow-dominant watersheds are projected to lead to increased flood risk (Salathé et al. 2014), which can stress infrastructure and create water quality problems.

Coastal parts of the Northwest are vulnerable to sea level rise, although projected impacts vary by locality. Some coastlines are expected to experience a relative decrease in sea level because tectonic uplift may outpace sea level rise, while other areas could see an increase in sea level of 50 inches or more (Dalton et al. 2013).

Overall, the impact of climate change Northwestern drinking water supplies will vary across the region. Systems which rely on snow for water supply will become more vulnerable. Water supplies from groundwater wells and aquifers may experience further declines (especially due to increasing withdrawals), although river basins that have a significant groundwater component may be somewhat buffered from climate change (Tague et al. 2013).

Alaska and Hawaii

Alaska has warmed twice as fast as the rest of the United States, leading to rapidly melting sea ice, shrinking glaciers, and thawing permafrost (Melillo et al. 2014). Summer sea ice in the Arctic is expected to disappear by the middle of the century. The decline in sea ice creates a positive feedback loop that generates more warming. Additionally, warming ocean temperatures are projected to increase ocean acidification and impacts on marine fisheries. Native communities are highly vulnerable to these changes, despite their history of adaptation to change (Melillo et al. 2014).

Hawaii and the US Affiliated Pacific Islands will also experience warmer ocean temperatures in the future, along with coastal flooding and erosion due to both sea level rise and high water levels from tropical and extra-tropical storms. Sea level rise is expected to threaten freshwater availability due to saltwater intrusion into coastal aquifers, especially in low-lying areas. Warmer future temperatures, decreased rainfall, and increased risk of drought will also reduce available water supplies for both drinking water and agriculture on most islands in the region (Melillo et al. 2014).

Canada

Canada's significant land mass experiences a range of climates that vary by region and season. Warming has been observed across the country, especially in the north and west, and less so in the east. In particular, daily minimum temperatures have increased faster than daily maximum temperatures over the last 60 years (Bush et al. 2014). Annual precipitation has increased by about 16 percent over the same time period, with most stations showing increased seasonal precipitation in spring and fall. Some sites in western Canada show a decrease in winter precipitation due to the decline in winter snowfall (Bush et al. 2014). Precipitation in Canada is greatly influenced by natural climate variations like ENSO and the PDO.

Climate model projections on average indicate that the nation is projected to warm by about 3–5°F (1.5–2.5°C) by the middle of the century, even in a low emissions scenario (Bush et al. 2014). Winter temperatures in particular are expected to increase by about 5–13°F (3–7°C) by the end of the century under a higher emissions scenario. Some regions like the High Arctic and Hudson Bay will experience much greater warming (Bush et al. 2014). Projected precipitation changes are more variable across models, although most of the country is projected to be wetter in all seasons, with the exception of southern Canada (Bush et al. 2014). Extreme heat days will occur more often, and the return period of extreme daily precipitation events is expected to become shorter (Bush et al. 2014).

Effects on Canada's water resources will depend on regional conditions. In river basins fed by glaciers, the loss of glacier ice is expected to reduce runoff, with potential long-term impacts on water availability for drinking water and hydropower generation (Bush et al. 2014). Similar trends in the rain–snow ratio will affect the timing of spring runoff and will decrease the amount of summer runoff in many river systems. Some coastal areas could see increases in sea level, especially during storm surges, leading to coastal erosion and other impacts (Bush et al. 2014).

Mexico

Mexico spans a range of climate conditions and ecosystems, from arid high plateaus to forested tropical zones. The country is expected to see a decrease in precipitation with climate change. Declines of up to 30 percent by 2040 are projected across many regions of the country, with the most acute declines projected in semi-arid northwestern Mexico (IPCC 2014), the primary region of irrigated grain farming. In areas of Mexico with significant subsistence agriculture, climate change scenarios suggest an increase in temperature from 2–4°F (1–2°C) and a reduction in precipitation by 10 percent, which may cause declining water availability as drought risk increases as well as increased exposure of rural populations and the agricultural sector (IPCC 2014). Although in some regions an increase in precipitation may contribute to greater rangeland productivity (IPCC 2014), the effects of maximum summer temperatures on livestock heat stress are expected to reach the level at which losses can occur by 2020 and will continue to rise (IPCC 2014).

While overall precipitation may decline in most areas, extreme precipitation events and powerful hurricanes may lead to higher flood risk. As a result, many of Mexico's agricultural communities are considered highly vulnerable, especially due to their high sensitivity and low adaptive capacity to these changes (IPCC 2014). Migration from Mexico to the United States has been associated with natural disaster occurrence and poverty trends (IPCC 2014), as well as with decreasing precipitation (IPCC 2014). Climate change may exacerbate these migration drivers.

ASSESSING WATER SYSTEM VULNERABILITY

Considering potential regional climate impacts is a necessary first step for water utilities, but impacts are strongly dependent on locality and the particular hydrologic system. Utilities must be familiar with how each system or watershed operates in historical and current climate conditions before projecting how climate change could affect the hydrologic regime. The following questions are intended to guide further consideration of climate change vulnerabilities.

- Is the water supply surface water, groundwater, or both? How could changes in annual or seasonal temperature or precipitation affect the system?
- Is the hydrology of the system dominated by rainfall or snowmelt? What are the implications in regions where the rain–snow ratio is expected to change?
- How sensitive historically is the hydrology and water quality of the system to changes in temperature, precipitation phase, or groundwater recharge rates?
- What extreme events has the water system experienced (floods, rainstorms, windstorms, wildfires, heat, droughts, storm surges)? How has the system responded?
- Are there any extreme event thresholds above which the system fails?
- Where are facilities and infrastructure located? Are they exposed to flooding, storm surges, sea level rise, or wildfires?
- How does customer water demand change based on temperature and precipitation?

Most water utilities have access to extensive data sets available from historical weather records, maps, stream gages, snow telemetry (SNOTEL) sites, tidal gauges, and other monitoring efforts. These data sets can be analyzed to help answer the previous questions, even if a utility has not used climate models to project future changes to its system.

Some studies have also analyzed paleoclimatology, which uses proxy data such as tree rings, ice cores, and pollen to reconstruct and extend the historical hydrologic and climatologic record. The paleo-record may indicate that a watershed or region has experienced extreme events such as long-term drought that predate the historical record. Paleoclimate or dendrohydrology information "can be used to define an expected range of variability, including worst-case scenarios, to guide operational strategies in the face of extreme future uncertainty" (Biondi and Strachan 2012). For example, tree-ring data used to reconstruct streamflows in the Upper Colorado River Basin indicated that streamflows vary over decades or longer and that short-term records may be inadequate for long-term planning (Woodhouse et al. 2006).

Resources on Regional Climate Impacts and Assessing Water System Vulnerability

This section provides summaries of publicly available (online) resources that provide an array of information, data sets, and interactive tools that water utilities can use to further understand potential regional climate impacts and to assess the vulnerability of hydrologic systems to climate change.

Adaptation Strategies Guide for Water Utilities, Climate Ready Water Utilities. This guide, prepared by the USEPA Climate Ready Water Utilities (CRWU) program, provides drinking water and wastewater utilities with a basic understanding of how climate change can impact utility operations and missions, as well as examples of actions utilities can take (i.e., adaptation options) to prepare for these impacts.

Canada in a Changing Climate: Sector Perspectives on Impacts and Adaptation. This report is a 2014 update to the 2008 science assessment report "From Impacts to Adaptation: Canada in a Changing Climate." Led by Natural Resources Canada, this report involved more than 90 authors and 115 expert reviewers and synthesized more than 1500 recent publications to update the state of climate science for Canada.

Changes in Water Use under Regional Climate Change Scenarios, Water Research Foundation. This research study by the Water Research Foundation sought to increase the adaptive capacity of water utilities in planning for and responding to climate change impacts, concentrating on the demand for water.

Climate Adaptation Knowledge Exchange (CAKE). CAKE was founded by EcoAdapt and Island Press in July 2010 and is managed by EcoAdapt. It aims to build a shared knowledge base for managing natural and built systems in the face of rapid climate change. CAKE helps users vet and organize the best information available. It also builds a community via an interactive online platform, creates a directory of practitioners to share knowledge and strategies, and identifies and explains data tools and information available from other sites.

Climate Change Handbook for Regional Water Planning. This handbook was developed as a partnership of the Environmental Protection Agency Region 9, the California Department of Water Resources, US Army Corps of Engineers South Pacific Division, and the Resources Legacy Fund. Although this handbook is focused on the California Integrated Regional Water Management Planning process, it can be used by other practitioners nationally and internationally when incorporating climate change into any watershed or water supply planning process. This handbook considers both climate change adaptation and mitigation.

Climate Change and Water Resources: A Primer for Municipal Water Providers. This primer was prepared by the National Center for Atmospheric Research and sponsored by the AWWA Research Foundation (AwwaRF) in 2006. The goals of this Primer are to (1) introduce water utility managers to the science of climate change; (2) suggest the types of impacts it can have on water resources; and (3) provide guidance on planning and adaptation strategies.

Climate Data Initiative. This US federal government data initiative provides data related to climate change that is related to coastal flooding, food resilience, water, ecosystem vulnerability, human health, energy infrastructure, transportation and the Arctic region to help inform and prepare America's communities, businesses, and citizens.

CREAT: Climate Resilience Evaluation and Awareness Tool. This risk assessment tool from the Environmental Protection Agency CRWU program allows water utilities to evaluate potential impacts of climate change and identify adaptation options to address impacts using both traditional risk assessment and scenario-based methods. Users are able to quantify the cost of climate hazards to utilities, with and without adaptation.

Implications of Climate Change for Urban Water Utilities. This report was prepared by Stratus Consulting for the Association of Metropolitan Water Agencies. The paper is designed to help urban water utilities move past the initial barrier of information overload and uncertainties related to climate change. The intent is to provide an essential understanding and then turn to consideration of the issues involved in developing suitable water sector responses to climate change.

National Climate Assessment. The US Global Change Research Program (USGCRP) Fourth National Climate Assessment was completed in 2018 and summarizes the impacts of climate change on the United States now and in the future. A team of more than 300 experts guided by a 60-member Federal Advisory Committee produced the report, which was extensively reviewed by the public and experts including federal agencies and a panel of the National Academy of Sciences. Readers can review climate science, regional impacts, sectoral impacts, and response strategies.

NOAA Regional Integrated Science Assessments. The US National Oceanic and Atmospheric Administration's Regional Integrated Sciences and Assessments (RISA) program supports research teams that help expand the nation's capacity to prepare for and adapt to climate variability and change. Central to the RISA approach are commitments to process, partnership, and trust building. There are currently nine funded RISAs across the United States that provide resources to water resource managers.

US Climate Resilience Toolkit. The US Climate Resilience Toolkit offers a process-based approach to initiate, plan, and implement projects to become more resilient to climate-related hazards and provides real-world case studies describing climate-related risks and opportunities. The toolkit also has a catalog of freely available tools to access and analyze climate data, generate visualizations, explore climate projections, estimate hazards, and engage stakeholders in resilience-building effort. The Water Resources section of the toolkit focuses on municipal water supplies, flooding, drought, and ecosystems.

US Department of Interior Climate Science Centers. The US Department of Interior Climate Science Centers (CSCs) partner with natural and cultural resource managers to provide science that helps ecosystems and the communities they support adapt to climate change.

REFERENCES

Abatzoglou, J.T., Rupp, D.E. and Mote, P.W. 2014. Seasonal Climate Variability and Change in the Pacific Northwest of the United States. *Journal of Climate.* 27:(5):2125–2142. doi: dx.doi.org/10.1175/JCLI-D-13-00218.1.

Alliance for the Great Lakes. Great Lake Issues: Conserving Water. Accessed Oct. 7, 2015. greatlakes.org.

Asefa, T., Adams, A. and Wanakule, N. 2015. A Level of Service Concept for Planning Future Water Supply Projects Under Probabilistic Demand and Supply Framework. *Journal of American Water Resources Association.* 51(5):1272–1285.

Biondi, F. and Strachan, S. 2012. *Dendrohydrology in 2050: Challenges and Opportunities in Toward a Sustainable Water Future: Visions for 2050.* (W.M. Grayman, D.P. Loucks, and L. Saito, edtiors). Sponsored by Emerging and Innovative Technology Committee and the Environmental and Water Resources Institute (EWRI) of the American Society of Civil Engineers. ascelibrary.org/doi/pdfplus/10.1061/9780784412077 (accessed Feb. 24, 2020).

Bush, E.J., Loder, J.W., James, T.S., Mortsch, L.D. and Cohen, S.J. 2014. An Overview of Canada's Changing Climate; in Canada in a Changing Climate: Sector Perspectives on Impacts and Adaptation, (ed.) F.J. Warren and D.S. Lemmen; Government of Canada, Ottawa, ON, p. 23-64.

Cayan, D., Kunkel, K., Castro, C., Gershunov, A., Barsugli, J., Ray, A., Overpeck, J., Anderson, M., Russell, J., Rajagopalan, B., Rangwala, I. and Duffy, P. 2013. Future Climate: Projected Average. *In Assessment of Climate Change in the Southwest United States: A Report Prepared for the National Climate Assessment* (G. Garfin, A. Jardine, R. Merideth, M. Black, and S. LeRoy, editors). Washington, D.C.: Island Press, 153–196.

CDM. 2011. *Climate Change Handbook for Regional Water Planning. Prepared for US Environmental Protection Agency Region 9 and California Department of Water Resources.* In partnership with US Army Corp of Engineers South Pacific Division, Resources Legacy Fund, and US

Environmental Protection Agency Office of Research and Development. www.esf.edu/glrc/library/documents/ClimateChangeHandbookforRegionalWaterPlanning_EPA_2011.pdf (accessed Feb. 24, 2020).

Cook, E.R., Seager, R., Heim, R.R., Vose, R.S., Herweijer, C. and Woodhouse, C. 2010. Megadroughts in North America: Placing IPCC Projections of Hydroclimatic Change in a Long-Term Palaeoclimate Context. *Journal of Quaternary Science*. 25:(1):48–61.

Dalton, M.M., Mote, P.W. and Snover, A.K. editors. 2013. Climate Change in the Northwest: Implications for our Landscapes, Waters and Communities. Washington, DC: Island Press. cig.uw.edu/news-and-events/publications/climate-change-in-the-northwest-implications-for-our-landscapes-waters-and-communities/ (accessed Feb. 24, 2020).

Dettinger, M. 2011. Climate Change, Atmospheric Rivers, and Floods in California—A Multimodel Analysis of Storm Frequency and Magnitude Changes. *Journal of the American Water Resources Association*. 47:(3)514–523, doi:10.1111/j.1752-1688.2011.00546.x.

Dettinger, M.D., Ralph, F.M., Das, T., Neiman, P.J. and Cayan, D.R. 2011. Atmospheric Rivers, Floods and the Water Resources of California. *Water*. 3:(2):445–478.

Fischbach, J.R., Lempert, R.J., Molina-Perez, E., Ahad Tariq, A., Finucane, M.L. and Hoss, F. 2015. *Managing Water Quality in the Face of Uncertainty: A Robust Decision Making Demonstration for EPA's National Water Program*. Santa Monica, Calif: RAND Corporation.

Fritze, H., Stewart, I.T. and Pebesma, E. 2011. Shifts in Western North American Snowmelt Runoff Regimes for the Recent Warm Decades. *Journal of Hydrometeorology*. 12:5:989–1006.

Frumhoff, P.C., McCarthy, J.J., Melillo, J.M., Moser, S.C. and Wuebbles., D.J. 2007. Confronting Climate Change in the U.S. Northeast: Science, Impacts, and Solutions. Synthesis report of the Northeast Climate Impacts Assessment (NECIA). Cambridge, MA: Union of Concerned Scientists (UCS).

Grayman, W.M., Loucks, D.P. and Saito, L. (editors). 2012. Toward a Sustainable Water Future: visions for 2050. Sponsored by Emerging and Innovative Technology Committee and the Environmental and Water Resources Institute (EWRI) of the American Society of Civil Engineers. ascelibrary.org/doi/pdfplus/10.1061/9780784412077.

Hayhoe, K., Wake, C.P., Huntington, T.G., Luo, L., Schwartz, M.D., Sheffield, J., Wood, E., Anderson, B., Bradbury, J., DeGaetano, A., Troy, T.J. and Wolfe, D. 2007. Past and Future Changes in Climate and Hydrological Indicators in the US Northeast. *Climate Dynamics*. 28:4:381–407.

Hidalgo, H.G., Das, T., Dettinger, M.D., Cayan, D.R., Pierce, D.W., Barnett, T.P., Bala, G., Mirin, A., Wood, A.W., Bonfils, C., Santer, B.D. and Nozawa, T. 2009. Detection and attribution of streamflow timing changes to climate change in the western United States. *Journal of Climate*. 22(13)3838–3855.

Hoerling, M.P., Dettinger, M., Wolter, K., Lukas, J., Eischeid, J., Nemani, R., Liebmann, B. and Kunkel. K.E. 2013. Present Weather and Climate: Evolving Conditions. In: *Assessment of Climate Change in the Southwest United States: A Report Prepared for the National Climate Assessment* (G. Garfin, A. Jardine, R. Merideth, M. Black, and S. LeRoy, editors). Washington, D.C.: Island Press: 74-97.

IPCC (Intergovernmental Panel on Climate Change). 2014a. Climate Change 2014: Synthesis Report. Contribution of Working Groups I, II and III to the Fifth Assessment Report of the Intergovernmental Panel on Climate Change [Core Writing Team, R.K. Pachauri and L.A. Meyer, editors]. Geneva, Switzerland; IPCC: 151 pp. www.ipcc.ch/pdf/assessment-report/ar5/syr/SYR_AR5_FINAL_full.pdf (accessed Jan. 15, 2020).

IPCC. 2014b. AR5 Climate Change: Impacts, Adaptation, and Vulnerability. Contribution of Working Group II to the Fifth Assessment Report (WGII AR5) of the Intergovernmental Panel on Climate Change. www.ipcc.ch/site/assets/uploads/2018/02/WGIIAR5-Chap26_FINAL.pdf (accessed Aug. 18, 2020).

IPCC. 2007. Climate Change 2007: Synthesis Report. Contribution of Working Groups I, II and III to the Fourth Assessment Report of the Intergovernmental Panel on Climate Change (Core Writing Team, Pachauri, R.K. and Reisinger, A., editors). Geneva, Switzerland: IPCC. www.ipcc.ch/site/assets/uploads/2018/02/ar4_syr_full_report.pdf (accessed March 30, 2020).

IPCC. 2001. Climate Change 2001: The Scientific Basis. Contribution of Working Group I to the Third Assessment Report of the Intergovernmental Panel on Climate Change [Houghton, J.T., Y. Ding, D.J. Griggs, M. Noguer, P.J. van der Linden, X. Dai, K. Maskell, and C.A. Johnson, editors]. Cambridge University Press, Cambridge, United Kingdom and New York, NY, USA, 881pp.

IPCC. 2001. Climate Change 2001: Synthesis Report. Contribution of Working Groups I, II, and III to the Third Assessment Report of the Intergovernmental Panel on Climate Change

IPCC. 2000. Special Report on Emissions Scenarios. (Nakicenovic, N., and R. Swart, eds.). Cambridge University IPCC. 2000. IPCC Special Report on Emissions Scenarios: Summary for Policymakers. A Special Report of IPCC Working Group III. www.ipcc.ch/pdf/special-reports/spm/sres-en.pdf.

Press, Cambridge, United Kingdom. www.ipcc.ch/report/emissions-scenarios/?idp=0.

Kunkel, K.E., Stevens, L.E., Stevens, S.E., Sun, L., Janssen, E., Wuebbles, D., Redmond, K.T. and Dobson, J.G. 2013a. Regional Climate Trends and Scenarios for the U.S. National Climate Assessment: Part 5. Climate of the Southwest U.S. NOAA Technical Report NESDIS 142-5. 87 pp., National Oceanic and Atmospheric Administration, National Environmental Satellite, Data, and Information Service, Washington, D.C.

Kunkel, K.E., Karl, T.R., Easterling, D.R., Redmond, K., Young, J., Yin, X. and Hennon, P. 2013b. Probable Maximum Precipitation and Climate Change. *Geophysical Research Letters*. 40:7:1402–1408.

Luce, C.H. and Holden, Z.A. 2009. Declining annual streamflow distributions in the Pacific Northwest United States, 1948-2006. Geophysical Research Letters, Vol. 36, L16401, doi:10.1029/2009GL039407.

Meko, D.M., Woodhouse, C.A., Baisan, C.A., Knight, T., Lukas, J.J., Hughes, M.K. and Salzer, M.W. 2007. Medieval drought in the upper Colorado River Basin. *Geophysical Research Letters*. 34, 10705, doi:1029/2007GL029988.

Melillo, J.M., Richmond, T.C. and Yohe, G.W. (editors). 2014. Climate Change Impacts in the United States: The Third National Climate Assessment. U.S. Global Change Research Program, 841 pp. doi:10.7930/J0Z31WJ2.

Milly, P.C.D., Betancourt, J., Falkenmark, M., Hirsch, R.M., Kundzewicz, Z.W., Lettenmaier, D.P. and Stouffer, R.J. 2008. Stationarity is Dead: Whither Water Management? *Science.* 319:5863: 573–574. DOI: 10.1126/science.1151915

Monterrosso, A., Conde, C., Gay, C., Gómez, D. and López, J. 2014. Two methods to assess vulnerability to climate change in the Mexican agricultural sector. *Mitigation and Adaptation Strategies for Global Change.* 19(4):445–461.

Mote, P.W. 2006. Climate-driven variability and trends in mountain snowpack in western North America. *Journal of Climate.* 19(23):6209–6220.

National Weather Service. Glossary. w1.weather.gov/glossary/index.php?letter=w (accessed Aug. 18, 2020).

National Weather Service. Climate Prediction Center. www.cpc.ncep.noaa.gov/products/precip/CWlink/MJO/enso.shtml (accessed March 30, 2020)

National Weather Service. JetStream Max: What a Cycle! www.weather.gov/jetstream/whatacycle_max (Accessed Jan. 15, 2020).

Nolin, A.W. and Daly, C. 2006. Notes and Correspondence: Mapping "At Risk" Snow in the Pacific Northwest. *Journal of Hydrometeorology*. 7(5):1164–1171.

Northrop, P.J. and Chandler, R.E. 2014. Quantifying Sources of Uncertainty in Projections of Future Climate. *American Meteorological Society*. 27(23):8793–8808.

Patz, J.A., Vavrus, S.J., Uejio, C.K. and McLellan, S.L. 2008. Climate change and waterborne disease risk in the Great Lakes region of the U.S. *American Journal of Preventive Medicine*. 35(5):451–458.

Routson, C.C., Woodhouse, C.A. and Overpeck, J.T. 2011. Second Century Megadrought in the Rio Grande Headwaters, Colorado: How Unusual Was Medieval Drought? *Geophysical Research Letters*. 38:22.

The Royal Society & National Academy of Sciences. 2014. Climate Change Evidence & Causes: An overview from the Royal Society and the US National Academy of Sciences. The Royal Society & National Academy of Sciences. royalsociety.org/-/media/Royal_Society_Content/policy/projects/climate-evidence-causes/climate-change-evidence-causes.pdf (accessed Oct. 8, 2020).

Rupp, D.E., Abatzoglou, J.T., Hegewisch, K.C. and Mote, P.W. 2013. Evaluation of CMIP5 20th Century Climate Simulations for the Pacific Northwest US. *Journal of Geophysical Research: Atmospheres*. 118(19):10,884–10,906.

Salathé, E.P., Hamlet, A. F., Jr., Mass, C. F., Lee, S., Stumbaugh, M. and Steed, R. 2014. Estimates of Twenty-First-Century Flood Risk in the Pacific Northwest Based on Regional Climate Model Simulations. *Journal of Hydrometeorology*. 15:1881–1899.

Stratus Consulting and Denver Water, 2015. Embracing Uncertainty. A Case Study Examination of How Climate Change Is Shifting Water Utility Planning. Prepared for the Water Utility Climate Alliance (WUCA), the American Water Works Association (AWWA), the Water Research Foundation (WRF), and the Association of Metropolitan Water Agencies (AMWA) by Stratus Consulting Inc., Boulder, CO (Karen Raucher and Robert Raucher) and Denver Water, Denver, CO (Laurna Kaatz). www.wucaonline.org/assets/pdf/pubs-uncertainty.pdf (accessed Oct. 8, 2020).

Tague, C.L., Choate, J.S. and Grant, G. 2013. Parameterizing sub-surface drainage with geology to improve modeling streamflow responses to climate in data limited environments. *Hydrology and Earth System Sciences*. 17(1):341–354.

USGCRP (US Global Change Research Program). 2018. Impacts, Risks, and Adaptation in the United States: Fourth National Climate Assessment, Volume II (Reidmiller, D.R., C.W. Avery, D.R. Easterling, K.E. Kunkel, K.L.M. Lewis, T.K. Maycock, and B.C. Stewart, editors. US Global Change Research Program, Washington, DC,. nca2018.globalchange.gov/downloads/NCA4_2018_FullReport.pdf (accessed Oct. 8, 2020).

US Drought Monitor. April 2019 map of California drought. The National Drought Mitigation Center. droughtmonitor.unl.edu/data/pdf/current/current_ca_trd.pdf (accessed Oct. 8, 2020).

USEPA (US Environmental Protection Agency). 2016. What Climate Change Means for New York. EPA 430-F-16-034. Washington, D.C.: USEPA.

Woodhouse, C.A., McKim, E. and Ray, A. 2006. New Streamflow Reconstructions for the Upper Colorado River Basin: Placing Recent Droughts into a Centuries-Long Context. Feature Article From Intermountain West Climate Summary, June 2006. wwa.colorado.edu/climate/iwcs/archive/IWCS_2006_June_feature.pdf (accessed Oct. 8, 2020).

WCRP (World Climate Research Programme). 2015. WCRP Coupled Model Intercomparison Project (CMIP). wcrp-climate.org/wgcm-cmip/about-cmip (accessed Oct. 12, 2015).

Zetland, D. 2014. Living with Water Scarcity. First edition. Aguanomics Press: Amsterdam, Mission Viejo, Vancouver. livingwithwaterscarcity.com.

Chapter 4

Principles of Engineering Practice

Past practices must be modified given the uncertainty, variability, and immediacy of climate change impacts. For a generation of engineers and utility operators who are used to designing infrastructure to a "Factor of Safety" or "Level of Service," operating and managing infrastructure in an uncertain climate requires a paradigm shift. This chapter reviews the principles of engineering practice, emphasizing that there is more than one way to apply engineering principles to utility management in light of uncertain future.

PAST PRACTICE

Most public and private water infrastructure currently in service was designed with the assumption of climate stationarity, the thought that past climate data are statistically representative of future climate. This assumption fails to account for long-term climate trends.

A probability density function can be fitted to past data, and well-established statistical methodologies can guide the selection of appropriate design criteria. An engineer may be interested in extreme design events (e.g., to accommodate peak design flows in new infrastructure) or in an average condition (e.g., snowpack for operational planning). However, developing statistically valid design criteria is often challenging due to sparse data records, especially for relatively long return periods. In some cases, it may be possible to supplement sparse data sets with data from similar sites with longer records or to use other climate modeling methods. Regardless, data sufficiency is challenging at many sites.

These challenges associated with an incomplete understanding of climate have been faced by engineers in the past, yet engineers have designed infrastructure around the world that meets its service objectives. Traditional concepts of engineering conservatism, combined with sound professional judgment, allow engineers to partially overcome incomplete knowledge of the past and uncertainty about the future.

DESIGNING FOR THE FUTURE

The climate has always been changing. Increasing scientific knowledge has led to a greater understanding of natural climate cycles such as the shorter-term El Nino–Southern Oscillation (ENSO) and the longer-term Pacific Decadal Oscillation (PDO). Depending on factors such as the time period of interest, these cycles may influence the design analysis.

Now that anthropogenic climate change is known to threaten infrastructure, design analysis must also account for a shifting climate. As described in Chapter 3, different modeling techniques project different future climate scenarios. All modeling is uncertain. An ensemble approach that considers a broad range of potential climate scenarios is often preferred.

While engineers have always considered uncertainties such as future population growth and regulations in their analysis and designs, it is now recognized that climate change poses extraordinary uncertainty. In some cases, the uncertainty may be so great that traditional conservative engineering approaches to bound uncertainty may be ineffective, yielding grossly over-conservative designs. As a result, engineers may need to carefully align level of service with actual need.

LEVEL OF SERVICE

Engineered infrastructure is normally designed for some specified level of service (LOS). Some works, such as dams, may be designed with extreme conditions in mind (e.g., safe passage of Probable Maximum Flood), with protection of human life as the paramount objective. Other infrastructure may be designed to address more common conditions, such as nuisance flooding, inadequate water pressure, or service outages. LOS may also be set by statute or regulation. For example, Florida Statute Chapter 373.709 (2013) sets the LOS requirement for water supply planning, stating that "the level-of-certainty planning goal associated with identifying the water supply needs of existing and future reasonable-beneficial uses must be based upon meeting those needs for a 1-in-10-year drought event." LOS is an implicit acknowledgment of uncertainty in the system. It provides the basis for what customers are willing to pay as well as what utilities are risking in order to provide a given LOS. There is a cost associated with moving from one LOS to the other. Customers' preference for a given LOS can be identified through a willingness to pay for an increased LOS or willingness to accept a lower LOS than the existing level (Howe et al. 1994).

In some cases, the desired level of service may be set by historical precedent or common practice and may not reflect the LOS that is actually required. The governing LOS for dams and other infrastructure with extreme failure consequence will often remain the same despite climate impacts. However, the LOS for other infrastructure may require reconsideration. There is evidence (New York City DEP) that some communities are willing to pay to reduce risk associated with major inconveniences such as drought, but the same principles may not apply for less consequential concerns (e.g., water pressure). Willingness to pay for improvements in LOS for a changing climate can be quantified informally through media (Seattle Times 2014) or via a formal targeted survey—the contingent valuation method (Howe et al. 1994).

While the LOS provided by infrastructure such as dams may remain the same in a changing climate, the return period of the design storm may change. For example, a 100-year return flood of "Q" quantity may become a more frequent occurrence. A spillway structure designed for that storm will still be able to provide protection for the same "Q."

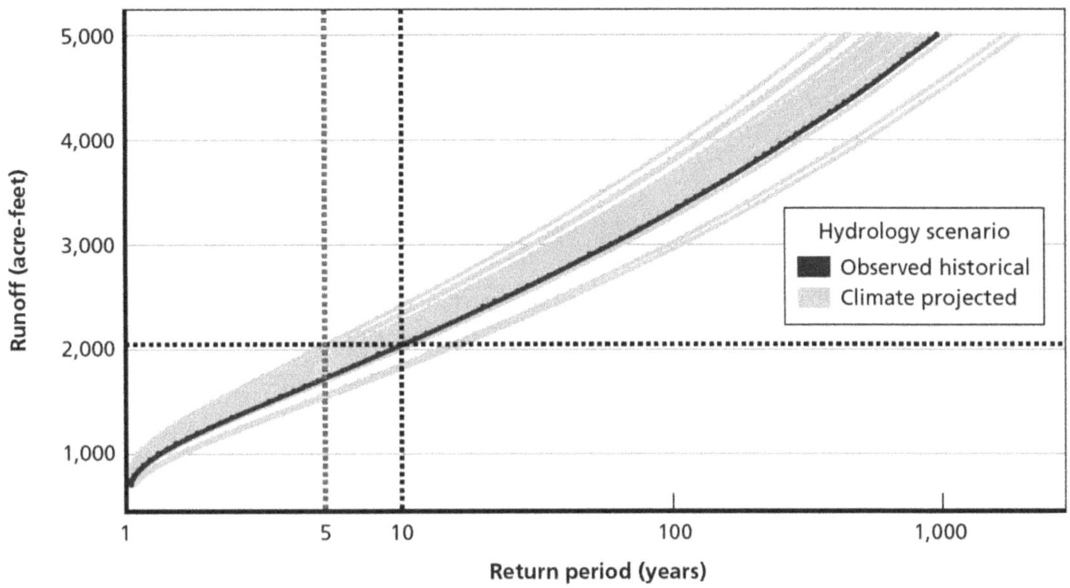

Source: Fischbach et al. 2015.

Figure 4-1 Observed historical and climate-altered hydrology daily runoff volume return periods for the Patuxent river basin

Note: Solid black line shows daily runoff by return period from the observed historical hydrology scenario. Gray shading shows the range of daily runoff projections from the climate projected hydrology scenario. Black dashed lines highlight the 10-year daily runoff, while the red dashed line shows how this volume might occur more frequently in one extreme climate projection.

However, an assessment must be performed to align expectations with the changing frequency of floods. For example, Figure 4-1 shows a 10-year return period flood that is projected to shift to a 5-year return period according to one climate model in Patuxent River Basin, Maryland.

Types of Level of Service

The following section includes a detailed discussion of the various types of LOS.

Emergency versus non-emergency scenarios. Emergency response may not be directly considered during design depending on the risk of consequence. For flooding associated with extreme weather, LOS might be reduced if aligned with improved emergency response protocols. For a water distribution main break, however, one has to take into account how the LOS will be met during an emergency response. Cost-benefit and consequence analyses may be required to balance infrastructure (capital) expenditure and emergency (operational) response.

Operational versus planning. The time horizon for services may guide how to assess changes in LOS in a changing climate. LOS for day-to-day operations such as water quality, pressure, and flow requirements are different from LOS for long-range efforts such as water supply planning (Asefa & Adams 2013). The LOS, and what it will require to maintain that LOS, should be clearly communicated to customers. For example, Seattle Public Utilities (SPU) incorporates LOS, life cycle costing, and triple bottom line criteria into their sewer maintenance program (Asefa et al. 2013). SPU sets a target and tracks the number of sewer backups per 100 miles of pipe. These data drive maintenance and replacement programs.

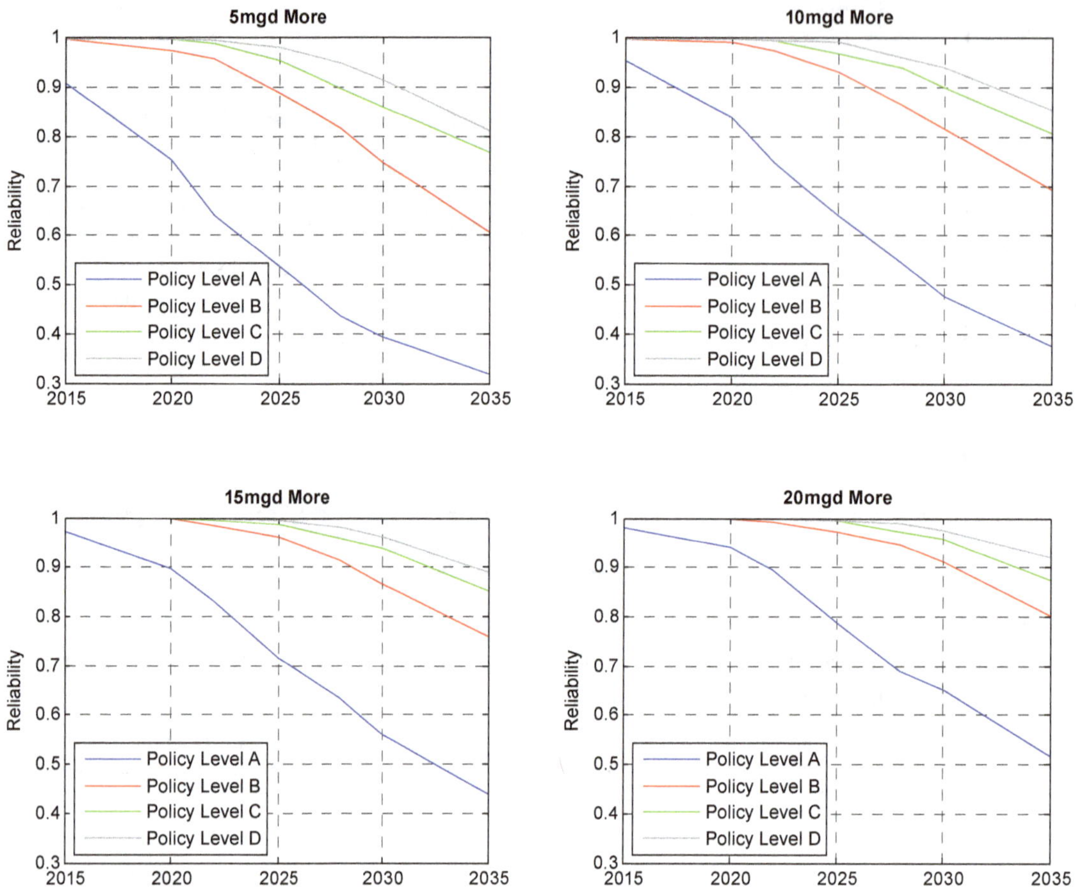

Source: Asefa et al. 2015

Figure 4-2 Improvement in level-of-service reliability through addition of incremental supply to meet projected demand

Note: Each policy level corresponds to the ability to meet demand considering different drought frequencies. For example, Policy level A represents meeting forecasted future demands without any intervention 99 percent of the time.

Tampa Bay Water incorporates the LOS approach into water supply planning efforts. Figure 4-2 illustrates how different LOS (Policy Levels A through D) may be improved by adding additional water supply sources at different times in the planning period. The LOS is tied to the frequency of drought and thus to the supply availability and demand predictions.

Deterministic versus probabilistic. Deterministic LOS is associated with a fixed number that can be communicated to customers as such—for example, SPU's number of sewer backups. In contrast, a probabilistic approach recognizes the uncertainty around any threshold and tries to estimate the chance of achieving it. By their nature, regulatory LOS tend to have a fixed threshold, such as the concentration of a constituent that is monitored by regulatory agencies, or the water quantity and pressure level that a wholesale customer needs at a specific point of delivery. Since providing an LOS costs money, it is important to recognize the uncertainty inherent in establishing a given LOS. Probability-based LOS try to address this uncertainty by estimating the reliability of achieving a given LOS. For example, Figure 4-2 shows that the reliability of meeting a Policy A LOS in 2035 with an additional 20 mgd source is about 50 percent, compared to 80 percent reliability for a Policy Level B LOS (allowing demand to be met without intervention 95 percent of the time). This comparison demonstrates the cost of meeting demand during extreme drought conditions.

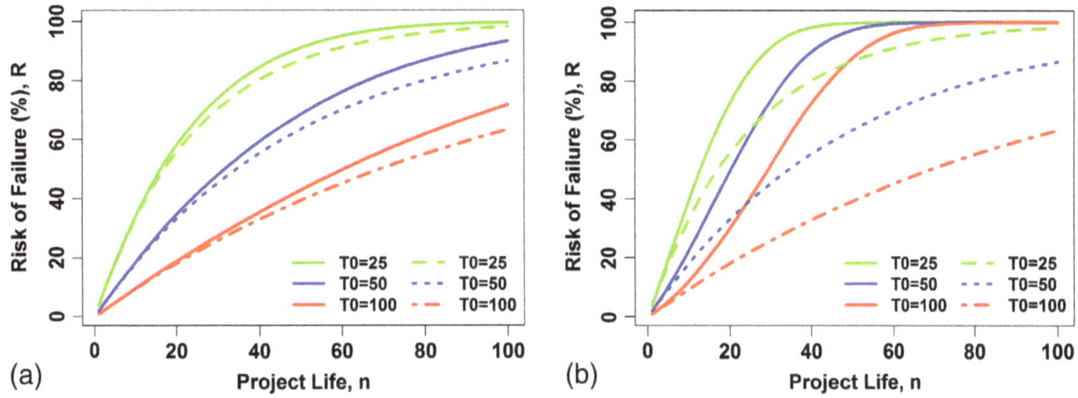

Source: Figure 8 from Revisiting the Concepts of Return Period and Risk for Nonstationary Hydrologic Extreme Events, *ASCE Journal of Hydrologic Engineering*, March 2014. With permission from ASCE

Figure 4-3 Non-stationary risk of failure as a function of project life, n

Note: Assuming for TO = 25, 50, 100 year; dashed lines show the risk for stationary condition, whereas the solid lines show the risk for nonstationary condition (a) Aberjona River Basin and (b) Little Sugar Creek Basin.

Level of Service in a Changing Climate

Climate change compounds existing system uncertainties, making it difficult to establish an LOS. A reliability-based, probabilistic perspective addresses these challenges systematically, making clear the cost associated with achieving a given LOS. For example, Figure 4-3 shows that most general circulation model (GCM) climate scenarios project a greater chance of getting an n-year return period flow more frequently (most projections are above the black historical line). However, there are climate scenarios that differ (grey lines below the historical curve), as shown in the figure. Once a range of climate scenarios are explored, one can estimate the probability of having, say, a 100-year return flow in 50 years or less by simply counting the number of climate scenarios that meet this requirement. For example, there is an 89 percent chance that a 10-year return daily flow becomes more frequent in a changing climate than what was observed historically.

Failure to provide a given LOS can be quantified using a nonstationary risk calculation approach, as in Figure 4-3. This approach differs from stationary risk calculation, where the underlying assumption is that the key statistics of a time series stay the same. The figure shows that the risk of failure increases across all project life spans when based on a nonstationary assumption.

CLIMATE CHANGE IMPACT ASSESSMENT FOR UTILITIES: BEST PRACTICES AND CHALLENGES

One of the first steps in the process of understanding the LOS for an entity is to understand the impacts from climate change. It is important to understand what a typical climate change impact assessment for a utility looks like. It is also important to understand the pitfalls and the best practices. Any adaptation measure and LOS determination effort that utilities undertake first requires an understanding of how climate change may impact their specific system. The best way to understand the impacts is to model them by means of climate models. Key factors must be considered before applying a GCM to a local area. Such studies have important implications; they may guide a costly capital improvement project to improve the adaptive capacity of a utility, for example, so utilities must take a measured approach to implementing adaptation measures. Several assumptions regarding local climate projection data are often ignored by the application community. The next sections describe the best practices and notes on the usage of the climate models to gather climate change–related information.

Prepare a Hydrological or Systems Model for Your Local Condition

A utility must understand its own local hydrologic condition before trying to determine what a changing climate could bring to that locality. The data, time, and space scales of the model dictate the time-space of locally-adjusted climate projection data that utilities should seek. A recent study conducted by the Water Utilities Climate Alliance (WUCA) followed four utilities' climate change impact assessments and found that one of the most important steps was having a well-calibrated model representing the utilities' hydrologic condition ready (WUCA 2015). It is also important to understand the uncertainties as well as the input and parameter sensitives in the local hydrological model as these may mask the range of possible climate scenarios that could be derived from the climate change projection. Without understanding what these models are predicting and synthesizing a specific decision that is relevant to a utility, climate scenarios may not provide much value. Instead, utilities should improve their hydrological and systems model before embarking on a climate change impact assessment. If climate change projections cannot be discriminated from the model uncertainty with confidence, it will be difficult to tease out the impact of climate change. This situation could require "tightening" the hydrological model. However, even well-calibrated models depend on the input data, so the applicability of the conceptual model embedded in these models must be checked carefully.

Ground Truthing Climate Projection Data

There is no guarantee that a climate model that successfully reproduced historical data will be able to accurately predict future conditions. Unfortunately, there is no other way to determine if a GCM projection is accurate. By design, downscaling and bias correction methods force GCMs to conform to what is observed locally, potentially masking how well a GCM reproduced local conditions. Before conducting these analyses, it is important to identify which of the GCMs are relevant to a utility's locality. Models are fine-tuned based on the modeler's familiarity with a region. A model that has high skill in Europe may not have high skill in North America. Similarly, a model that suits the Pacific Northwest may not be suitable in the Southeast. One must acknowledge the differences in GCMs' skills between different locations. Ground truthing also requires the availability as well as space–time scale of the variable of interest in selecting which of the GCMs should be tested. Given that there are hundreds of GCMs available, one must know the conditions driving these GCMs.

Dynamical versus Statistical versus Hybrid Downscaling

GCMs are usually run at a high spatial scale such as 200 km and are too coarse for direct input into a utility's model. A statistical or dynamical downscaling model can translate this large-scale climate output into local-scale hydrologic input. Statistical downscaling methods rely on first developing a transfer function between GCM outputs and hydrological model inputs observed at local scale. The transfer function could be a simple interpolation from the center of a GCM cell to the local scale, or a sophisticated regression—such as a multi-regression with exogenous variables, neural nets, and support vector machines—that relates GCM outputs to local-scale input parameters. Because of its huge computation cost savings, statistical downscaling is more popular than dynamical downscaling in impact assessment. Some GCM runs can easily be downscaled in a matter of minutes, depending on the scale and complexity of the project. These models assume that the statistical relationship built from historical data and historical GCM runs (known as retrospective runs) is stationary in a changing climate. In addition, the physical processes that influence local climate at a much smaller scale, such as sea breeze or Bermuda High affecting Southeast United States precipitation patterns, may not be captured by these models. Their use depends on the kind of decision to be made based on the impact study, particularly whether local processes are relevant at the scale of the decision.

Unlike statistical downscaling techniques, dynamical downscaling techniques (known as regional climate models or RCMs) embed the climate physics that drive local climate conditions. They use large-scale GCM output as a boundary condition to drive the finer-scale, physics-based model through time. The accuracy of a dynamical model regarding a climate signal that is not in the parent GCM model (especially low-frequency signals such as ENSO or PDO) remains a research question. However, it is clear that the local physics driving climatic conditions for an area are better captured in a dynamically downscaled product than via a statistical approach. In addition, less bias correction is necessary than for many statistical models (see the section titled, "The Change Factor Method versus Bias Correction" later in this chapter). The disadvantage of these models is their huge computational cost, which makes it cost-prohibitive to run multiple scenarios for a given region. Utilities must decide whether a few dynamical downscaled products would be more useful than a range of GCM scenarios analyzed using cost-effective statistical downscaling.

Recently, researchers have created hybrid downscaling techniques to provide a compromise between these two methods. Hybrid downscaling techniques rely on building the transfer function between the GCM and RCM (dynamical downscaled data) as opposed to the statistical downscaling models that derive transfer functions by using GCM and observed data. Hybrid downscaling techniques use enough GCM to RCM relationships to establish the transfer function, then takes this established relationship to downscale several other GCMs, saving the computational cost of an RCM. Ideally, the transfer function is "trained" on dynamically downscaled data that embed the local physics and thus have a better representation of the local climate. While this approach makes intuitive sense, the benefit of this technique has yet to be demonstrated in a wide range of applications.

Top-Down versus Bottom-Up

There are two ways of incorporating climate change information into an impact assessment. The most widely used approach is usually referred to in literature as a top-down approach or a chain of model approach (WUCA 2015), indicating the direction of the flow of information during the process. A top-down climate impact assessment usually starts with few selected GCM ensembles (see the section titled, "Ensemble Projection: To cull or to Not Cull?" later in this chapter) either by downscaling the parameters of interest such as temperature or precipitation for a given locality or by using the change factor information (called the change factor approach or delta-method). These downscaled data then pass through hydrological operational models, providing key decision variables of interest that the utility can use to develop an adaptation action. This process is to be repeated for all GCMs selected, giving a range of decision variables over which probability distribution is defined. Expected values are then calculated based on the weight given to each GCM. If all GCMs are considered likely (equal weight), the expected value is simply the average over the sample size. This approach is sometimes called "predict, then act."

In contrast, the bottom-up approach starts by identifying the decision point that is important for a utility and defining the vulnerability domain that one wants to investigate in a changing climate. This results in a climate response function relating key decision variables with a set of climate variables whose range of change would be informed by

GCM projections. For example, based on historical data, one could perform a range of sensitivity analysis for meeting projected demand in 2040 as a function of precipitation and temperature changes. The analysis is not projecting what would happen in 2040; instead, it gives the quantity of shortage a utility would potentially face for a given change in precipitation and temperature. Once this relationship between decision and climate variable is defined, GCM projections are used to derive the likelihood of a shortage and whether adaptation is necessary. An advantage of this approach is that for some cases it is not necessary to rerun the hydrologic and systems model using all of the GCM outputs. Instead, the GCMs can be mapped on the climate response function to derive the likelihood over which a key decision must be made.

While the bottom-up approach (also known as decision scaling or climate-informed decision making) has gained popularity over the last few years, it is still limited by the likelihood function derived from the GCM, which may not fully capture the range of possible future outcomes—the very problem top-down approaches are criticized for. The method also relies on neatly mapping the climate response function through sensitivity of climate input variables (such as precipitation and temperature changes). Climate response mapping is also a many-to-one function. Figure 4-4 illustrates the top-down and bottom-up approaches.

The Change Factor Method versus Bias Correction

Because representing local-scale physical processes is difficult, outputs from large-scale GCMs have bias that impedes their direct use in hydrological impact analysis (Wood et al. 2002, Li et al. 2010). Downscaling addresses the spatial mismatch by producing finer-scale

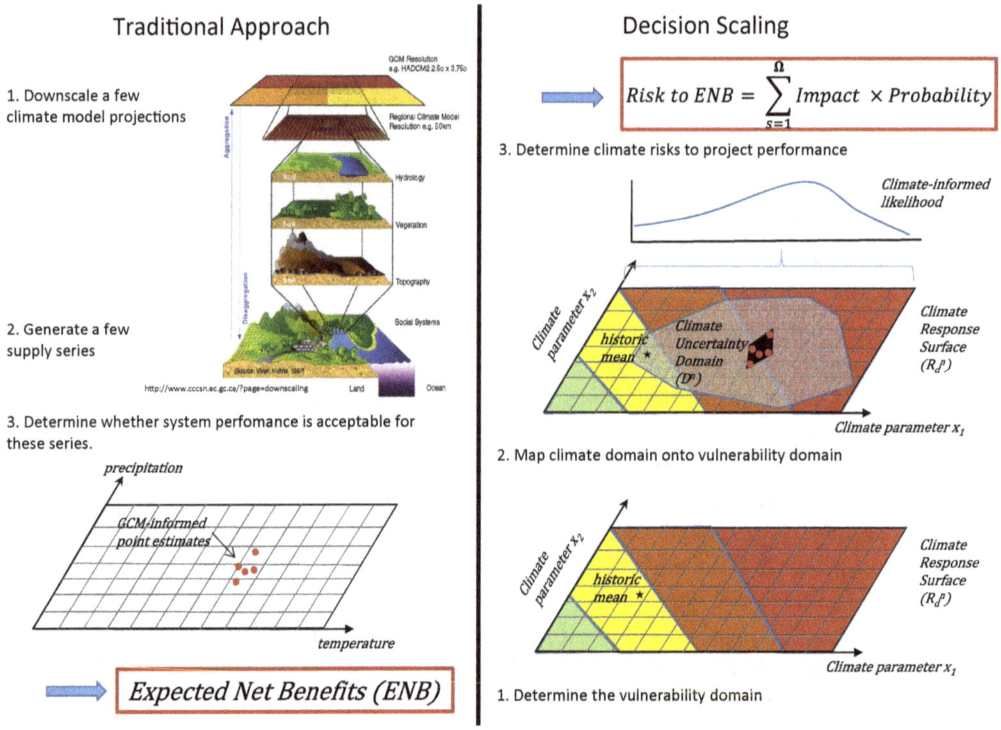

Source: Adapted from Garcia et al. 2014

Figure 4-4 An example of top-down and bottom-up climate change impact assessment framework

GCM outputs. In contrast, bias correction injects local scale variability into an otherwise smooth GCM output. Some researchers hold that rather than "correcting" GCMs to match historical observation, the change in climate predicted between historical and future simulations should be used. This approach is known as the change factor method or delta method. In its simplest form, the change factor method uses the changes predicted by a GCM—say an x percent decrease for precipitation and a y percent increase in temperature over 30 years—and applies that change to historical data of similar length to generate a future hydrological time series. These data then feed into a systems model that informs a future climate change impact.

One concern with this approach is that it assumes the same change factor for an entire distribution of parameters of interest. Changes in mean behavior and changes for an extreme case would be presented as similar, when in reality this may not be the case. Anandhi et al. (2011) present a sophisticated change factor method that derives the change factors at different percentiles of the parameters. One limitation of the technique is that the time sequence of future GCM outputs is not used; rather, this method uses the change from the historical time period that is being modeled.

A widely used bias correction technique known as quantile mapping (Panofsky and Brier 1968, Wood et al. 2002) relies on matching the probability distribution of retrospective GCM runs with historical observation. This method yields modified GCM retrospective data with the exact probability distribution, while the time sequence usually differs. While "snapping" the GCM retrospective runs on historical data, the mismatch creates a bias that is then used to "correct" future GCM projections, resulting in an adjusted future time series to be used by a hydrological model.

Important Time Scales in Bias Correction Methods for Precipitation

Currently, several GCMs have sub-daily climate parameter outputs, typically three to six hours. While the existence of fine time scale GCM output is exciting, GCM skill is limited at this fine scale. It is not common practice to use sub-daily time scale in the incorporation of GCM data. Daily or monthly time scales are preferred. Correcting a daily precipitation time series for bias presents an additional challenge because of no-rain days. Should the modeler use the historical no-rain-days probability and modify the probability of the amounts only when generating future scenarios? Should the number of no-rain days that are output from the GCMs be corrected? These questions must be answered before using any correction techniques described in the previous section for daily time scales. If the data are provided to the modeler through an internet data provider portal, it is important to understand the assumption that is considered in creating the data. To avoid this dilemma, some experts recommend using monthly precipitation data and historical resampling (known as analog technique) to project daily precipitation values. While this approach avoids the subjective decision-making process involved with no-rain days, it also ignores the daily sequence data informed by a changing climate scenario. The decision should depend on the objective of the impact study. Are daily time sequences important for this objective? On the other hand, one must consider the derivation of intensity-duration-frequency, which relies on sub-daily time-steps.

Cautionary Note on Bias Correction

Bias correction usually matches historical probability distribution segregated by month for both daily and monthly data. There is no preservation of time series characteristics (daily or monthly transition). In fact, while one could perfectly match the daily or monthly probability density function of historical observation, the time sequence of what the GCM produced and what was actually observed can be completely out of sync. Test whether important statistical characteristics, such as transition probability at multi-time scales, are reasonably reproduced by retrospective runs. It is important to understand this limitation while making decisions. GCMs cannot effectively reproduce large-scale persistent atmospheric signals such as ENSO and AMO. For example, a warm ENSO phase produces consistent above-average rainfall to the lower half of the United States, and this cycle repeats every 4 to 8 years. We still do not know how the frequency-amplitude of this signal is impacted by a changing climate.

Ensemble Projection: To Cull or to Not Cull?

In time series modeling, ensemble forecasting is a well-tested approach that is proven to be successful. Ensemble forecasting usually outperforms a single model prediction, especially if the ensembles are "trained" based on different components of the time series. GCM output for a given locality usually differs for the same emission scenario or even the same model. There are many sources of uncertainty in these forecasts (Knutti et al. 2010), including: (1) uncertainty in the process being modeled, that is, a poor understanding of the process in the first place; (2) uncertainty in model parameters, and (3) structural uncertainty as a result of the inability to describe a known process accurately. Because of these, each GCM will have a different level of uncertainty reproducing relevant parameters such as precipitation and temperature for a given region.

It is widely recognized that output from an ensemble of models is more robust than output from a single GCM (Brekke et al. 2008, Knutti et al. 2010). While most agree that practitioners should use an ensemble of GCMs, there is no consensus on whether one should use some form of weighting mechanism to do so. Using any weighting mechanism means weight for each GCM will be equal and the decision would be made based on all the GCMs. Those in favor of weighting hold that, like any other model, GCMs should be vetted based on how well they reproduce historical data before giving them credibility. Those who oppose model culling argue that each GCM simulation provides one possible scenario of the future and that there is no way to know which GCM provides a better projection. In addition, superior past performance does not necessarily guarantee better future performance (Knutti et al., 2010). Weighting GCMs depends on the metrics used to test their performance (Brekke et al. 2008).

One of the most widely used approaches in weighting GCMs was proposed by Giorgi and Mearns (2002). This method is based on a reliability ensemble average technique that defines two criteria of bias and convergence. The method was formalized by Tebaldi et al. (2005) in a Bayesian framework. The method weights a GCM based on how well it reproduced historical data (bias criterion) and how close it is to the ensemble mean for future scenario (convergence criterion). Sometimes this approach is termed the truth-centered approach. In this approach, retrospective GCM runs are assumed to be sampled from a distribution that is centered on observation plus some error, while future projections are centered on the mean of GCM ensembles (Asefa and Adams 2013). Regardless of the specific criterion used to weight GCMs, it is important to identify the parameters of interest for a given locality. Is it precipitation that drives the GCM selection, or temperature, or the joint probability distribution?

Regional filters can also help select GCMs. For example, the California Department of Water Resources Climate Change Technical Advisory Group suggests an approach based on the use of different filters, starting with global climatology filters, then the western US climatology and hydrology filter, then the California hydrology and extreme filter. The method was demonstrated to recommend 10 GCMs out of 31 initial pools (Lynn et al. 2015).

It is important to consider the multiple approaches to downscaling climate data and understand the inherent limitations associated with the various approaches before blindly adopting the downscaled data for utility-focused climate analyses.

PRACTICAL PLANNING STRATEGIES FOR INFRASTRUCTURE

Engineers should use the best available science and methods, as described in Chapters 3 and 4, to understand future climate drivers and implications for infrastructure design and operation. Engineers must continue to analyze and design infrastructure in the face of significant uncertainty about what our future climate will be. Sound professional judgment is required to properly interpret information and make appropriate decisions.

Existing Infrastructure—Short Remaining Service Life

Existing infrastructure, even if not resilient to the anticipated impacts of climate change, often still holds significant value. For existing infrastructure with a relatively short remaining service life, perhaps of fewer than 10 years, risk from climate change may be immaterial from a risk management perspective. This infrastructure may continue to be used safely, perhaps complemented with parallel efforts at improved emergency response or diverting loading. Infrastructure may be improved with either capital upgrades, improved maintenance, or operational enhancements. In other cases, particularly if the infrastructure is already being used past its intended design life, complete replacement of the infrastructure may be required.

Existing Infrastructure—Long Remaining Service Life

Existing infrastructure with a relatively long remaining service life, perhaps a decade or more, will often not be fully resilient to the anticipated impacts of climate change. Vulnerability assessments will often prioritize the replacement or modification of infrastructure. However, risk can be manageable in some cases with increased emphasis on emergency response or operational enhancement.

Planning New Infrastructure

Except in unusual circumstances where new infrastructure is being designed for a very short service life, proposed new infrastructure should be analyzed and designed with future climates in mind. In some cases, the full required future capacity can be implemented in stages as required, such as securing land for future raising of a levee. It becomes increasingly important for engineers to document their assumptions as well as what conditions might cause an incremental enhancement to existing infrastructure.

Role of Engineering Judgment

Even when using sound analytical techniques and the best available science, sound professional judgement remains necessary, especially when extra conservatism may be appropriate. Traditionally, engineers often include additional safety measures to account for various uncertainties. Even with an incomplete knowledge of future climates, engineers can design and plan infrastructure that functions throughout its service life.

Role of decision-makers. Engineers are rarely the final decision-makers for major infrastructure decisions; typically, this role falls to elected officials and those who may not have a background in engineering techniques, infrastructure, or climate science. As the climate changes, engineering managers must engage with their elected officials to ensure a full understanding of the trade-offs between various options, especially for options with reduced initial capital costs but higher long-term commitments for improved emergency management, maintenance costs, and risks.

Application of supportive tools. A broad range of supportive tools are available to help practitioners evaluate the vulnerability of infrastructure select adaptive options. These tools include USEPA's Climate Resilience Evaluation and Awareness Tool (CREAT) and Engineers Canada's Public Infrastructure Engineering Vulnerability Committee (PIEVC) Protocol. A full description of the use and application of these tools is not included here, but is available from other sources, such as AWWA's Climate Change webpage (AWWA, n.d.). While all tools have their strengths and weaknesses, they can provide valuable support to the application of sound professional judgment.

REFERENCES

Anandhi, A., Frei, A., Pierson, D.C., Schneiderman, E.M., Zion, M.S., Lounsbury, D. and Matonse, A.H. 2011. Examination of Change Factor Methodologies for Climate Change Impact Assessment. *Water Resources Research*. 47:(3):W03501.

Asefa T. and Adams, A. 2013. Reducing Bias-Corrected Precipitation Projections Uncertainties: A Bayesian-Based Indicator-Weighting Approach. *Journal of Regional Environmental Change*. 13:(S1):111–120.

Asefa, T., Adams, A., and Wanakule, N. 2015. A level of service concept for planning future water supply projects under probabilistic demand and supply framework. *Journal of American Water Resources Association*. 51:(5):1272–1285.

AWWA (American Water Works Association). n.d. Climate Change. How Can We Prepare for the Impact of Climate Variability? www.awwa.org/Resources-Tools/Resources/Climate-Change (accessed Feb. 25, 2020).

Brekke, L.D., Dettinger, M.D., Maurer, E.P. and Anderson, M. 2008. Significance of Model Credibility in Estimating Climate Projection Distributions for Regional Hydroclimatological Risk Assessments. *Climatic Change*. 89:(3–4):371–394.

Fischbach, J.R., Lempert, R.J., Molina-Perez, E., Ahad Tariq, A., Finucane, M.L. and Hoss, F. 2015. Managing Water Quality in the Face of Uncertainty: A Robust Decision Making Demonstration for EPA's National Water Program. Santa Monica, CA: RAND Corporation.

Florida Statutes. 2013. Title XXVIII Natural Resources; Conservation, Reclamation, and Use. 373.709. Regional Water Supply Planning. www.leg.state.fl.us/Statutes/index.cfm?mode=View%20Statutes&SubMenu=1&App_mode=Display_Statute&Search_String=water+supply&URL=0300-0399/0373/Sections/0373.709.html (accessed Feb. 25, 2020).Garcia, L.E., J.H. Matthews, D.J. Rodriguez, M. Wijnen, K.N. DiFrancesco, and P. Ray. 2014. Beyond Downscaling: A Bottom-Up Approach to Climate Adaptation for Water Resources Management. AGWA Report 01. Washington, DC: World Bank Group.

Giorgi, F. and Mearns, L.O. 2002. Calculation of Average, Uncertainty Range, and Reliability of Regional Climate Changes from AOGCM Simulations via the 'Reliability Ensemble Averaging' (REA) Method. *Journal of Climate*. 15(10):1141–1158.

Howe, C.W. and Smith, L.O. 1994. The Value of Water Supply Reliability in Urban Water Systems. *Journal of Environmental Economics and Management*. 26:1:19–30.

Knutti, R., Furrer, R., Tebaldi, C., Cermak, J. and Meehl, G.A. 2010. Challenges in Combining Projections From Multiple Climate Models. *Journal of Climate*. 23:10:2739–2758.

Li, H., Sheffield, J. and Wood, E.F. 2010. Bias correction of monthly precipitation and temperature fields from Intergovernmental panel on climate change AR4 models using equidistant quantile matching. *Journal of Geophysical Research*. 115:(D10).

Lynn, E., Schwarz, A., Anderson, J. and Correa, M. 2015. Perspective and Guidance for Climate Change Analysis. California Department of Water Resources. Climate Change Technical Advisory Group. water.ca.gov/LegacyFiles/climatechange/docs/2015/Perspectives_Guidance_Climate_Change_Analysis.pdf (accessed Feb. 25, 2020).

Panofsky, H.A. and Brier, G.W. 1968. *Some Application of Statistics to Meteorology*. University Park: Pennsylvania State University, College of Earth and Mineral Sciences.

Salas, J.D. and Obeysekera, J. 2014. Revisiting the Concepts of Return Period and Risk for Nonstationary Hydrologic Extreme Events. *Journal of Hydrologic Engineering*. 19(3):554–568.

Seattle Times. 2014. https://www.seattletimes.com/nation-world/worldwide-rallies-call-for-action-now-on-climate-change/

Tebaldi, C., Smith, R.W., Nychka, D. and Mearns, L.O. 2005. Quantifying Uncertainty in Projections of Regional Climate Change: A Bayesian Approach to the Analysis of Multimodel Ensembles. *Journal of Climate*. 18:(10):1524–1540.

Wood, A.W., Maurer, E.P., Kumar, A. and Lettenmaier, D. 2002. Long-Range Experimental Hydrologic Forecasting for the Eastern United States. *Journal of Geophysical Research*. 107:(D20).

WUCA (Water Utilities Climate Alliance). 2015. Actionable Science in Practice: Co-producing Climate Change Information for Water Utilities Vulnerability Assessment. Las Vegas: WUCA.

This page intentionally blank.

Chapter 5

What Is a Climate Action Plan?

The previous chapters have laid the groundwork for this manual, addressing the purpose and intended audience, the science of climate change, and relevant engineering principles. This chapter addresses the key components of a climate action plan (CAP).

As noted previously, this manual focuses primarily on adaptation to climate change. Mitigation measures are addressed in other resources.

INTRODUCTION—ESTABLISHING THE CONTEXT

A CAP for a water or wastewater system prepares the system for the effects of climate change, particularly global warming, sea level rise, and related climatological and meteorological events. A CAP is not a vulnerability assessment, emergency response plan, energy optimization plan, nor a capital investment plan, although it works hand-in-hand with these plans. A CAP should include background information about climate change, as well as the threats relevant to water and wastewater systems.

The CAP introduction should include system-specific information, such as

- nearby water levels and elevations of facilities vulnerable to inundation,
- specific local raw water threats such as saltwater intrusion, and
- increased disinfection byproduct (DBP) challenges.

The introduction or a subsequent section should review research and local risk assessments that have been conducted.

The core sections of a CAP for a water or wastewater system should include

- impacts that must be addressed;
- measures and actions to be taken; and

- a climate risk assessment, using the US Environmental Protection Agency's (USEPA's) Climate Resilience Evaluation and Awareness Tool (CREAT) or comparable assessment process.

IMPACTS WATER AND WASTEWATER SYSTEM CAPS MUST ADDRESS

This section discusses both the direct physical impacts and the indirect impacts that a CAP for a water or wastewater system needs to include. *The impacts could have either a positive or a negative impact on the system.*

Direct Physical Impacts

Severe weather swings. Climate change will cause more severe swings in the weather, including more serious floods, droughts, and heat waves. These events will impact water and wastewater systems with greater severity than in the past.

Sea level rise. Sea levels will rise intermittently and permanently, inundating and damaging facilities near the water. By the very nature of water and wastewater systems, many facilities and infrastructure are near waterways or at a low point to make use of gravity (specifically, wastewater systems), including

- raw water pump stations,
- water treatment plants,
- sewer collection and transmission lines, and
- wastewater and resource recovery treatment plants.

Saltwater intrusion. Rising sea levels are expected to cause saltwater intrusion into groundwater and surface water supplies in many coastal areas. This will be especially likely in areas in which groundwater aquifers have already been excessively drawn down.

Diminished raw water supplies. Diminished surface and groundwater water supplies are expected due to less rain, less snow, earlier snowmelt, and greater evaporation.

More overflows. Severe storm events, rising water levels, and rising water table levels will cause more sanitary sewer overflows (SSOs) and combined sewer overflows (CSOs).

Indirect Impacts

Increased water demands. Water demands will increase, at least at times and more so in some locations than others, due to greater use of water for cooling and greater irrigation due to drought.

Forest fire runoff. Forest fires are expected to be more frequent and severe. This will cause raw water quality impacts from the associated fire runoff, which can contain contaminants, toxins, and very heavy sediment loading.

Increased algal blooms and toxins. Increased water temperature is expected to cause more frequent and severe blooms of algae as well as toxins (such as cyanotoxins).

Increased invasive and nuisance species. Increases in raw water temperature are expected to cause increases in invasive and nuisance species, such as the poisonous lionfish. Increased temperatures will provide longer growing seasons and reproductive seasons for invasive aquatic plants and animals, including zebra and quagga mussels, increasing their impact on the aquatic environment. Some invasive species can cause human illnesses and greater water treatment challenges.

Disinfection byproduct challenges. Increased raw water temperature increases the formation rate and maximum potential concentrations of natural organic matter (NOM) and DBP precursors. This will create water treatment and distribution challenges.

Greater sediment in raw water. Lower raw water intakes and more severe rains and forest fires will cause greater sediment in raw water, requiring greater treatment actions and greater wear on treatment equipment.

Water treatment impacts. Rising raw water temperatures will impact water and wastewater treatment processes, particularly those that use coagulants and oxidants.

Wastewater treatment impacts. Increased temperatures will impact biological wastewater treatment processes, potentially changing bacteriological populations and reactions. Bacterial activity doubles every 50°F (10°C). In some cases, the increased temperatures may benefit biological treatment processes, and wastewater treatment system operators must understand those effects. It may be possible to operate with lower populations of bacteria. This also may make direct potable reuse more attractive as raw water quality deteriorates and resource recovery effluents improve in quality.

Electrical power interruptions. More frequent and prolonged interruptions in electrical power supplies should be expected due to the impacts of severe weather on electrical transmission and distribution systems and the impacts of drought on water supplies used for power generation.

MEASURES AND ACTIONS CAPS MUST INCLUDE

This section is divided into the following subsections:

- Overarching and Preventive Measures
- Emergency Response Measures
- Operational and Programmatic Adaptation Measures
- Physical Adaptation Measures
- Other Measures

The most valuable adaptation measures are those that pose "no regrets" or offer "dual benefits." "No regrets" measures provide valuable benefits even if the climate does not change at all. "Dual benefits" measures provide benefits for purposes other than climate change. An example of this kind of adaptation measure is emergency electrical generators, which can greatly aid water and wastewater systems in a wide range of incidents. Utilities should consult resources such as USEPA's Adaptation Strategies Guide (USEPA, n.d.) or Georgetown Climate Center's Adaptation Clearinghouse to learn more about potential climate impacts specific to their location and possible adaptation measures.

Overarching and Preventive Measures

Establish a climate action planning team. To begin a CAP, water and wastewater systems must establish an internal climate action planning team. Depending on the agency's resources, this team may range from one person to many. Varying levels of staff should have input on the final product. Operators and mechanics bring important perspectives to discussion of impacts through their knowledge of existing systems.

The agency should also reach out to key external stakeholders, including critical customers such as hospitals. The forum for this external stakeholder engagement may be part of an existing organizational relationship or may require a new committee. Tools like United States Environmental Protection Agency's Workshop Planner for Climate Change and Extreme Events Adaptation provide materials to aid discussions with stakeholders. Stakeholder engagement is further discussed in Chapter 8 of this manual.

Stay informed. The CAP should include provisions for the climate action planning team and other agency personnel to learn more about environmental concerns, particularly climate change awareness, monitoring, and action triggers identification. Staying informed means being aware of climate projections, monitoring, and modeling. Water and wastewater CAPs need to address anticipated erratic precipitation patterns, including

- heavier rain events, along with longer and more frequent periods of less rain;
- more rain and less snow in areas of historic snowfall;
- more severe and unpredictable weather events;
- greater swings in precipitation levels; and
- more severe droughts.

Participate in regional efforts. Climate change affects all sectors and all locations. Water and wastewater systems should participate in regional collaborative efforts to benefit from collective efforts and to serve as one voice in the face of policy development.

Maintain the CAP. The CAP should include provisions to ensure it will be updated as climate science advances. Relevant local changes to address include

- changing financial resources;
- new or different local climate change planning organizations;
- new, modified, or closed facilities;
- new major customers or existing customers who have significantly reduced their water consumption; and
- new technologies that may lower adaptation costs.

Budget for CAP implementation. Each CAP must ensure that adequate funding is provided to execute the identified actions and documentation, from initial planning measures to long-term capital measures.

Emergency response planning. Regardless of preparation for climate change, water and wastewater systems should continue their existing emergency preparedness measures (refer to AWWA M19, *Emergency Planning for Water and Wastewater Utilities*), including

- dam emergency action plans (EAPs);
- vulnerability assessments (VAs);
- emergency response plans (ERPs), sometimes referred to as emergency action plans (EAPs);
- emergency operations plans (EOPs), or comprehensive emergency management plans (CEMPs);
- business continuity plans (BCPs), sometimes called continuity of operations plans (COOPs); and
- crisis communications plans (CCPs).

All of these plans should be regularly updated and integrated with the CAP.

Operational and Programmatic Adaptation Measures

Adapt water treatment processes. Water treatment processes may need to be adapted to address changing water conditions, including

- forest fire run-off,
- increased algal blooms and toxins,
- increased invasive and nuisance species,
- disinfection byproduct challenges,
- greater sediment in raw water, and
- increased water temperature.

Process adaptation may include the following.

- Coagulants are particularly sensitive to changes in water quality and temperature. Water systems may need to investigate alternative coagulants, since existing coagulants may not work efficiently as water quality changes. The investigation may result in seasonal usage of different coagulants.
- Different oxidants may be necessary for raw water pretreatment.
- Changes in coagulants, oxidants, or other treatment chemicals may require greater chemical storage requirements, either for multiple chemicals or for greater quantities of existing chemicals.
- Each raw water source will need to be individually monitored for qualitative and quantitative changes in the water due to climate change, with increased monitoring during drought and flood events. For impounded surface water sources, analytical samples should be taken routinely from various levels in the water body to monitor impacts on the raw water and treatability. Supervisory control and data acquisition (SCADA) system and laboratory data should be used to create specific reports based on raw water and finished water data as they relate to changing climate conditions. Most SCADA systems collect a tremendous amount of data that can be used more efficiently for process adaptation.
- Raw water should be monitored on an on-going on-line basis for parameters such as total organic carbon (TOC) and turbidity.
- A database of weather-related events should be maintained, as well as treatment data, to provide for correlating events with process changes.

Adapt wastewater treatment processes. Wastewater treatment processes may need to be adapted to address changes in influent wastewater, including increased temperatures and changing physical characteristics and bacteriological populations.

Develop drought response plans. Water utilities should develop formal drought response plans that are incorporated into or addressed in their CAP. Additional information is available in AWWA M60, *Drought Preparedness and Response Manual.*

These drought response plans should include

- reduced yield projections;
- clear trigger and conservation levels, excluding voluntary ones;
- consideration of reaction times;
- coverage of all types of customers;
- planning for "epic" drought conditions (i.e., worse than any existing drought severity levels);
- public education;
- water demand reduction tools;
- alternate raw water sources;
- system interconnections; and
- drastic measures.

Implement conservation measures. Detailed information on conservation planning and conservation measures is available in AWWA M52, *Water Conservation Programs—A Planning Manual.* In addition to special conservation measures during drought conditions, water systems should develop ongoing conservation programs that include

- public education;
- discouraging the use of potable water for irrigation, especially for non-native species; and

- water rates that encourage conservation, such as increasing block rates, and avoidance of rates that encourage excessive consumption, such as declining block rates.

Address interdependencies. As part of a CAP, agencies should identify interdependencies and address their roles in climate change preparedness. Interdependencies are sometimes also referred to as dependencies. Examples include the water sector's dependency on chemicals and the healthcare sector's dependency on water.

Maintain flexibility in Operations and Maintenance (O&M) programs. Water and wastewater systems need to maximize flexibility in O&M programs to be better able to respond to changing or new conditions that may arise due to climate change.

Physical Adaptation Measures

Establish emergency power. Water and wastewater systems are dependent on electrical power. As weather events and emergencies can impact the electrical power supply, water and wastewater systems need to invest in or have contracts in place for emergency electrical supplies such as permanent and/or portable generators. Generators must be maintained to ensure they will work when needed. Water and wastewater systems should also consider alternate (renewable) power opportunities, such as cogeneration systems at wastewater treatment facilities and solar panels at treatment plants that have available land and roof areas.

Adapt to sea level rise. Water and wastewater systems should address sea level rise in the planning, construction, and renovation of facilities, including water and wastewater treatment plants, pump stations, and sewer collection systems. By their nature, many water and wastewater facilities are close to water levels. Rising groundwater levels may "float" existing infrastructure, particularly gravity sewer mains. Since infrastructure can last for 50 to 100 years, serious consideration must be given to projected sea levels for that time period.

Elevate critical facilities and equipment. Special attention should be paid to highly vulnerable facilities and equipment, particularly electrical components of treatment plants and pump stations. In many cases, raising selected pieces of equipment can be a cost-effective adaptive measure.

Adjust water intakes. Raw water intakes should be adjusted for use in periods of drought, which usually requires installing lower intakes and relocating intakes horizontally to access a deeper portion of the body of water. Sometimes, completely new intake facilities may be necessary.

Construct greater raw water storage. More man-made storage will help protect source water from nuisance species, sedimentation, and forest fire runoff. The additional reservoirs can be filled when water quality is good and drawn down when raw water quality is impaired.

Install levees and sea walls. In some cases, facilities can be adequately protected by new or enhanced levees and sea walls. Levees and sea walls prevent the climate change-related events impacting the utility infrastructure. In the past, the levees and sea walls were designed for a future that was stationary but newer designs must incorporate the resiliency required to address the climate change-related uncertainty.

Relocate some facilities. Some facilities may need to be completely relocated. While this involves a substantial capital investment, it may be a reasonable alternative, particularly when major improvements are needed or new facilities are being constructed.

Establish alternate (diverse) sources of supply. Identifying alternate sources of supply is an excellent precaution. It may involve the following actions:

- Drilling additional wells in a groundwater system

- Adding surface water sources to a groundwater system
- Adding groundwater sources to a surface water system
- Recycled water sources
- New impoundments and/or intake points in an existing watershed
- New impoundments and/or intake points in an additional watershed

Develop emergency interconnections. Weather emergencies, even drought, may impact some systems more than others. Although they do not expand regional water availability, emergency interconnections between water systems are invaluable during disasters and water system emergencies. Interconnections should be developed to provide water flow in both directions, as large as is feasible, and regularly tested and exercised.

Increase the use of reclaimed water. Although increasing the production of reclaimed water does not increase the available water in a watershed, it allows the water to be used again before it is released downstream. With enough research and development, as well as public education and legislative authorization, direct potable reuse can become a mitigative solution to raw water shortages. Additional information on reclaimed water use is available in AWWA M24, *Planning for the Distribution of Reclaimed Water*.

Modify wastewater collection systems. Wastewater collection systems should be modified to minimize inflow and infiltration as much as possible and to accommodate high flows through improvements such as equalization facilities and basins. This will help prevent SSOs and CSOs, as well as help treatment plants deal with large flow swings from precipitation events.

Use green infrastructure. Green infrastructure uses either natural systems or engineered systems that mimic natural processes to control runoff and reduce water demand. Implementing green infrastructure improves stormwater collection and treatment, limits or prevents floods from peak storm flows into surface waters, augments groundwater supplies in shallow aquifers, and supports source water protection efforts. Projected changes in precipitation patterns from climate change will exacerbate existing challenges. Green infrastructure can be particularly effective when considered as part of a suite of options for stormwater management. For example, in conjunction with water conservation and infiltration/inflow reduction programs, it can be highly effective in reducing flow volumes. The use of green infrastructure has the potential to improve climate change response in addition to mitigation. For example, green roofs reduce the effects of rising temperatures and increasing rainfalls, while vegetative buffers slow and pretreat storm runoff and provide protection from winds and storm surge.

Provide public education. As part of a CAP, water and wastewater systems should educate employees, customers, and stakeholders. While broader public education can be provided by a number of agencies, the water or wastewater system should provide education regarding the impacts of climate change on the system and the planned adaptive measures. This topic is further addressed in Chapter 10.

AVAILABLE TOOLS FOR CAP DEVELOPMENT FOR WATER AND WASTEWATER SYSTEMS

Developing an effective CAP requires ongoing conversations with stakeholders and refining inputs. Utilities should revisit their CAP if knowledge of climate change impacts for their location changes, if they have implemented additional strategies that would potentially reduce risks, or if emerging technology reduces the costs of implementing potential adaptive measures. Even if the climate does not change as projected, utilities will still benefit from implementing "no regrets" strategies.

Water and wastewater systems nationwide may use existing tools to assist in the iterative process of developing CAPs. One of the leading climate risk assessment tools is

USEPA's CREAT; it is described in more detail in the subsequent section along with several other tools and guides.

USEPA's Climate Resilience Evaluation and Awareness Tool

CREAT (creat.epa.gov) was developed under USEPA's CRWU initiative in consultation with a working group that provided feedback on features and functionality. The working group consisted of representatives from drinking water and wastewater utilities, water sector associations, climate science experts, risk assessment experts, and federal agency partners.

The goals of CREAT are to

- increase drinking water, wastewater, and stormwater operator awareness of potential climate change impacts on utility operations and missions;
- help utilities determine threshold levels for asset failures and resulting consequences of an asset's inability to perform its designed function;
- guide utilities through the risk assessment process to quantify potential consequences from climate-related or other threats;
- inform adaptation decision-making by identifying and considering adaptation options that address identified threats and reduce associated impacts; and
- examine the cost of these different adaptation options in comparison to the economic losses associated with the consequences of climate change threats.

CREAT guides utility users through a series of intuitive steps to develop an understanding of climate threats, impacted assets, and adaptation options that will help mitigate those threats and protect service delivery. While CREAT does not provide an exhaustive list of all adaptive measures, impacts, and potential risks, the tool does act as a systematic first step to help utilities better understand risk and the potential monetized risk reduction gained by implementing adaptation measures.

CREAT begins with a review of climate science and climate change impacts. Users identify the location for their assessment, as well as basic utility information, such as population served, total flow, and financial condition. This condition indicates a utility's strength to endure operating revenue loss or capacity to expend funds to repair and replace equipment.

To identify potential climate-related risk, utilities must first recognize known concerns that are currently being addressed. Users identify these concerns, which help organize information to identify climate change threats, as well as assets to consider during the assessment. To explore and assess their current risk, users establish a *Baseline Scenario* based on data that are used for planning decisions and other assessments, such as historical data provided within CREAT or from records kept by the user. CREAT is flexible in its approach, and users can easily override the provided data with data from their region. The *Baseline Scenario* is compared to other scenarios in the Risk Assessment module or is used to help identify "no regrets" options. Users then consider CREAT-provided climate data by developing scenarios that represent a range of possible future climate conditions and the potential threats these conditions could generate. Based on the current concerns identified earlier in the process, CREAT provides default threat selections for the user's review. This is depicted in Figure 5-1.

Next, users are guided by CREAT to define the potential economic, environmental, and public health consequences of their climate threats on critical assets. Users define the consequences that could occur if a critical asset were to be destroyed, damaged, or rendered inoperable using an economic consequences matrix provided in CREAT to help make decisions systematically. Users then develop asset/threat pairs to assess the potential climate impacts on assets. An asset/threat pair is the unit of analysis for a climate change risk assessment; the focus is on the consequences to the critical asset if the threat were to occur across a number of scenarios.

As discussed earlier in the chapter, a better understanding of potential risks allows users to define potential adaptive measures and adaptation plans that a utility can use to protect their assets and mitigate the impacts of threats. These measures include those already implemented that provide resilience now, as well as potential measures that could increase resilience when implemented as part of adaptation plans. CREAT provides a number of potential adaptation measures and costs of implementation within the tool, but users are also able to input their own or consult other resources such as USEPA's Adaptation Strategies Guide or Georgetown Climate Center's Adaptation Clearinghouse for more ideas.

Finally, users can compare the potential costs and benefits of implementing adaptation measures in the face of projected climate threats as shown in Figure 5-2. CREAT provides monetized risk from assessments to support adaptation planning decisions

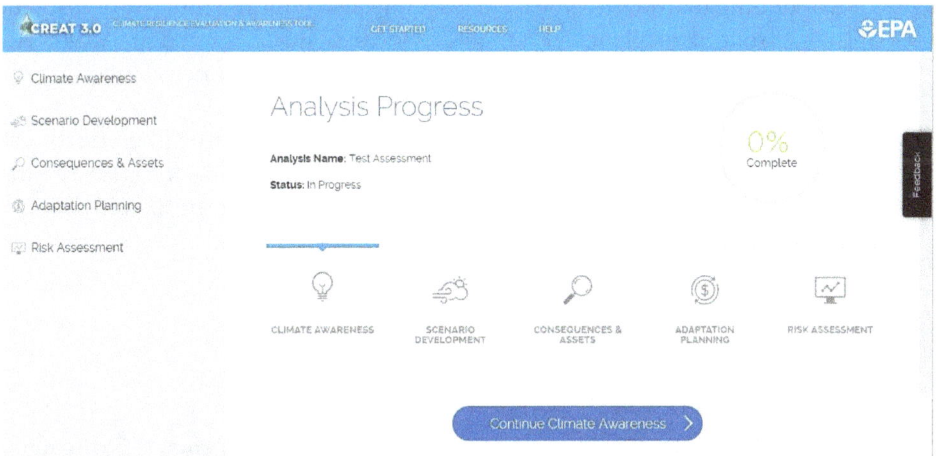

Source: Image created using the USEPA CREAT Tool Interface

Figure 5-1 USEPA's CREAT interface

Source: Image created using the USEPA CREAT Tool Interface

Figure 5-2 Comparison of various adaptation strategies developed using USEPA's CREAT

and characterize current and potential risks to utility assets. Monetized risk refers to the anticipated financial impact of a threat if it occurs, which is based on the consequences assessed for each critical asset. Results generated from this assessment can be compared in terms of their cost and potential risk reduction to identify the most effective approach. As with all steps in this risk assessment process, CREAT provides users with downloadable reports to inform adaptation planning or to determine if there is a need for further assessment.

For more information related to the climate risk assessment process and the climate science and economic data used in CREAT, users should access the CREAT 3.0 Methodology Guide. The Guide covers these topics in detail. Users can access the Guide directly from the CRWU homepage (www.epa.gov/crwu) or from the "Help" tab within CREAT.

USEPA's Drought Response and Recovery Guide

USEPA's Drought Guide (www.epa.gov/waterutilityresponse/drought-response-and-recovery-guide-water-utilities) provides worksheets, best practices, videos, and resources for responding to drought emergencies while also building long-term resilience. For water utilities, drought resilience is the ability to respond to immediate water supply threats, withstand drought impacts, and recover quickly. Recovery includes considering long-term conditions and planning for permanent solutions.

The Drought Guide was developed using information gathered from seven utilities with the help of a Water Sector Focus Group that relayed lessons already put into practice by their peers. The Drought Guide is divided into four sections to help users systematically plan for long-term resilience:

- Staffing, Response Plans, and Funding
- Water Supply and Demand Management
- Communication and Partnerships
- Case Studies and Videos

The Drought Guide uses an interactive PDF format, featuring a GIS-based Geoplatform, to create a compelling and engaging experience for users. Users are also provided with videos, links to outside resources, and downloadable worksheets to record their response and recovery planning progress.

USEPA's Workshop Planner for Climate Change and Extreme Events Adaptation

The Workshop Planner (workshopplanner.epa.gov/) was designed to be used by drinking water and wastewater utility staff, technical assistance providers, community leaders, and other stakeholders interested in conducting climate change adaptation workshops, with the goal of helping utilities and communities explore how intense and frequent extreme weather events can affect water resources.

The Workshop Planner guides users through a four-step process to plan and conduct a customized workshop centered on a user-selected climate threat scenario:

- Get Started—users enter basic workshop information, select extreme events and impacts to consider, and generate drafts of key workshop materials.
- Plan Workshop—users identify and contact workshop planning team members, refine workshop materials, engage in outreach to potential participants, and manage logistics.

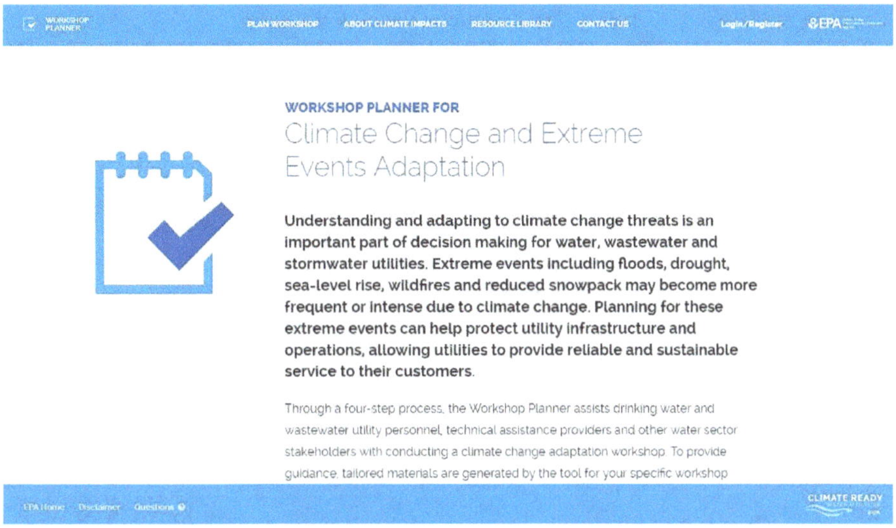

Source: Image created using the USEPA CREAT Tool Interface

Figure 5-3 User interface of USEPA's workshop planner for climate change and extreme events adaptation

- Final Workshop Preparation—users review guidance for facilitating and documenting the workshop and generate additional workshop handouts and materials to use in the workshop.
- Next Steps and Action Planning—users write a workshop report and implement action items identified during the workshop.

Users can explore how intense and frequent extreme weather events may affect water resources and can generate materials tailored to specific user needs. This information is presented in Figure 5-3. The Workshop Planner also provides helpful guidance for hosting a successful meeting—users of the tool will receive customized invitations, agendas, and presentations, as well as access to a built-in registration management system that allows users to invite and track participants.

USEPA's Adaptation Strategies Guide for Water Utilities

The Adaptation Strategies Guide (www.epa.gov/crwu/learn-how-plan-extreme-weather-events) provides drinking water and wastewater utilities with a basic understanding of how climate change can impact utility operations and missions and gives examples of different actions utilities can take to prepare for these impacts. Using four different types of informational briefs, the Guide walks users through an understanding of climate information specific to their location, what challenges the user might expect to see, and the adaptation options users can employ to address each climate challenge, as shown in Figure 5-4. An Adaptation Planning Worksheet is also included in the Guide to help identify and organize adaptation options that interest the user.

After reading the introductory material of the Guide, users can choose one of nine different US climate regions from a map based on their location. Selecting a region will bring users to the corresponding Climate Region Brief, which lists and describes projected changes in climate. Users are also pointed toward Group Briefs that provide more general information on regional climate impacts that many utilities face, including drought, flood, water quality degradation, ecosystem changes, and service demand and use.

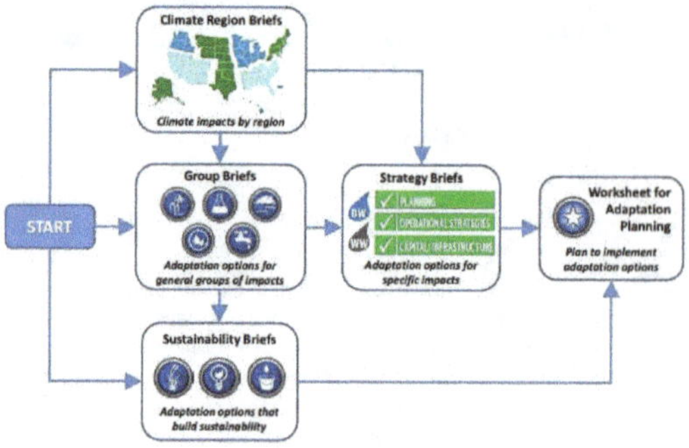

Source: Image created using the USEPA CREAT Tool Interface

Figure 5-4 Various information briefs presented in the USEPA's adaptation strategies guide for water utilities

With a better understanding of potential climate impacts, Strategy Briefs provide more detailed information on a particular impact and list adaptation options that, when implemented at a utility, can help to address that impact and ensure that water and wastewater systems are better prepared for related impacts. The Guide contains relative cost information related to each adaptation option as well. Finally, users can access Sustainability Briefs that describe how sustainable strategies—specifically, Green Infrastructure, Energy Management, and Water Demand Management activities—can be used in conjunction with adaptation planning.

CONCLUSION

Climate action plans are widely needed but are far from one-size-fits-all. This chapter introduces potential components that each water and wastewater system should consider for inclusion in their CAP, as well as publicly available resources that provide background and guidance on addressing impacts of climate change. The chapters that follow will address the "why" and "how" of CAPs.

REFERENCES

Adaptation Clearinghouse. www.adaptationclearinghouse.org/sectors/water/ Accessed: April 3, 2020

USEPA (United States Environmental Protection Agency). EPA Web Application Access wamssoprd.epa.gov/oam/server/auth_cred_submit?request_id=8426965752793794295&authmethod=FORM Accessed: April 3, 2020.

USEPA. Drought Response and Recovery: A Basic Guide for Water Utilities (41 pp, 12MB, August 2018, 810-B-17-003). Retrieved: April 3, 2020 from www.epa.gov/waterutilityresponse/drought-response-and-recovery-guide-water-utilities.

USEPA. CREAT Risk Assessment Application for Water Utilities www.epa.gov/crwu/creat-risk-assessment-application-water-utilities. Accessed: April 3, 2020

USEPA. Creating resilient Water Utilities (CRWU) www.epa.gov/crwu/ Accessed: April 3, 2020.

USEPA. Resilient Strategies Guide for Water Utilities www.epa.gov/crwu/resilient-strategies-guide-water-utilities#/ Accessed: April 3, 2020

Chapter 6

Why Do Utilities Need a Climate Action Plan?

This chapter discusses the need for utilities to develop a climate action plan (CAP) and the various drivers that compel a utility to consider creating one. Utilities operating without a CAP will face severe challenges in a future world driven by increasing uncertainty and variability due to a changing climate.

KEY DRIVERS FOR DEVELOPING A CAP

Trends Indicate Increase in Costly Extreme Weather Events

The increasing severity and frequency of damaging weather events shown in news headlines create a strain on emergency response budgets. Water practitioners today face the challenge of building resilience into current and future water infrastructure and system operations. As discussed in Chapter 3, the severity of extreme weather events has increased, impacting nearly every part of North America. These impacts include

- longer and more pronounced droughts in California, Texas, and the southwest;
- melting permafrost in the arctic;
- rising seas and storm surge along our coasts; and
- more frequent and extreme storms across North America that bring damaging winds, intense rainfall, and hazardous flooding.

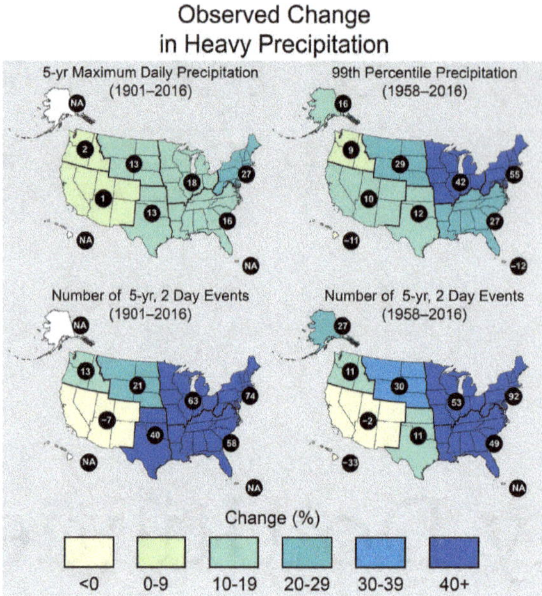

Source: CICS-NC and NOAA NCEI

Figure 6-1 Observed Change in Very Heavy Precipitation (1958–2012)

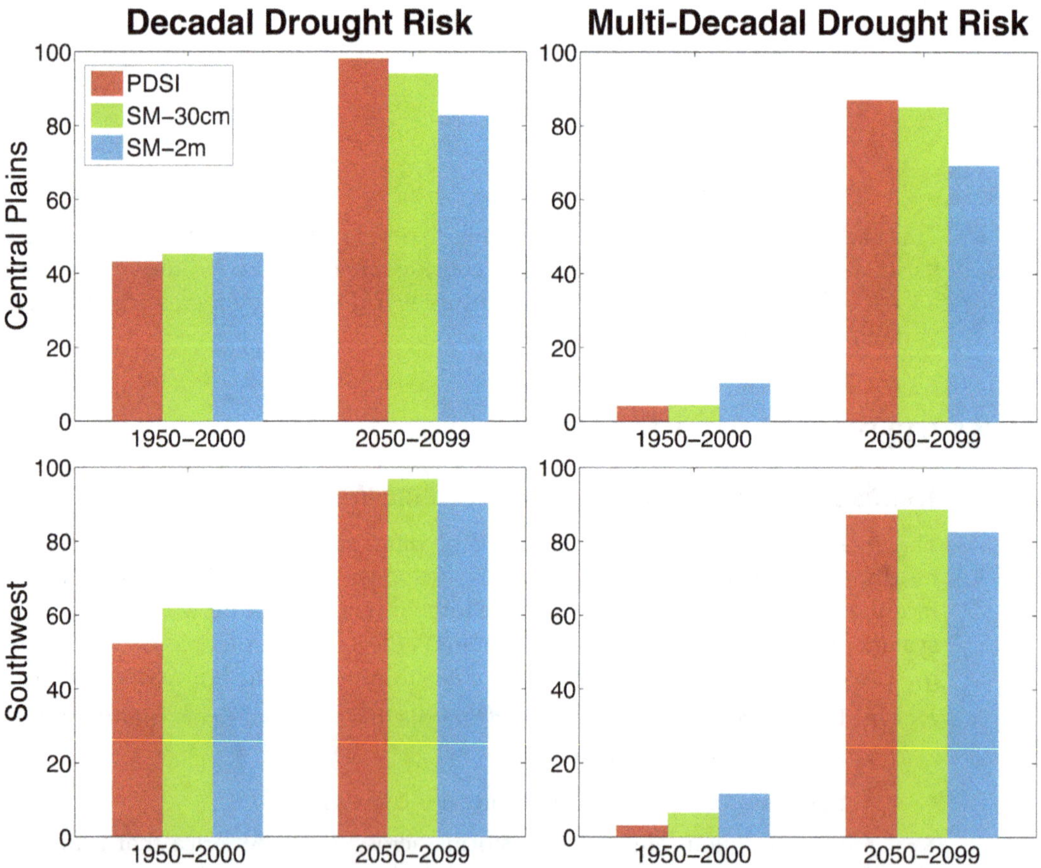

Source: Cook et al. Sci Adv. 2015; 12 February 2015

Figure 6-2 Risk (percent chance of occurrence) of decadal (11-year) and multidecadal (35-year) drought

The 2018 National Climate Assessment report has documented that, in the past 50 years, the heaviest one percent of storms has gotten more intense in all regions of the United States except Hawaii, as shown in Figure 6-1. At the other end of the extreme, the southwestern United States and Texas have experienced one of the worst droughts on record. Climate scientists are projecting that this could become the new normal, as shown in Figure 6-2. Paleoclimate reconstructions of drought and precipitation for California indicate that, though 3-year periods of persistent below-average soil moisture are not uncommon, the 2013 to 2015 drought event was the most severe drought in the last 1,200 years, with accumulated moisture deficits worse than any previous continuous span of dry years (Griffin and Anchukaitis 2015).

Drought not only diminishes water supplies but also increases forest mortality linked to an increasing threat of wildfires and infestations such as bark beetle (Williams et al. 2015). Diminished forest cover contributes to increased risk of erosion and turbidity, impacting surface water quality of reservoirs. Furthermore, the end of the severe drought was marked with severe flooding in both Texas and California in 2015.

Signals Indicate Near-Term Risks Induced by Climate Change Are Likely Emerging Now

In some regions, the projected effects of climate change are already occurring. The number of catastrophic storms continues to mount across the United States. Both large and small water utilities are being affected. In some cases, climate change manifests in the increased frequency of damaging flood events. For example, the community of Berwick, Maine (population 1,600), experienced two 100-year storm events that flooded their water treatment plant in consecutive years—2006 and 2007—a statistically unlikely set of events. A partial list of severe storms in the last decade follows.

- Washington, D.C., June 2006: 287 mm (11.3 in.) in 6 days
- Chicago, September 2008: 169 mm (6.64 in.) in 1 day
- Louisville, Ky., September 2009: 127 mm (5 in.) in 90 minutes, with rainfall rates up to 8.80 in./hour (224 mm/h)
- Atlanta, September 2009: 330 mm (13 in.) in 1 day
- Providence, R.I., March 2010: 345 mm (13.6 in.) in 30 days
- Nashville, Tenn., May 2010: 142 mm (5.6 in.) in 6 hours, 345 mm (13.6 in.) in 2 days
- Front Range (Colorado Springs to Fort Collins) of Colorado: September 2013: 230 mm (9 in.) in 1 day, 432 mm (17 in.) in 3 days
- Tampa, Fla., July 2015: 457 mm (18 in.) in 10 days
- Columbia, S.C., October 2015: 198 mm (7.8 in.) in 1 day, 450 mm (17.7 in.) in 5 days
- Austin, Tex., October 2015: 147 mm (5.8 in.) of rain in 1 hour, 394 mm (15.5 in.) in 1 day

Note: Calculated from the multimodel ensemble for Palmer Drought Severity Index (PDSI) and for soil moisture to 30 cm, and from 2 to 3 m (SM-30cm and SM-2m). Risk calculations are conducted for two separate model intervals: 1950–2000 (historical scenario) and 2050–2099 (RCP 8.5). Results for the Central Plains are in the top row, and those for the Southwest are in the bottom row.

Different Extremes Bring Different Challenges

It is important to understand what these changes mean for utilities. Figure 6-3 lists the range of climate change impacts projected for the water sector.

Source: Confronting Climate Change: An Early Analysis of Water and Wastewater Adaptation Costs, prepared by CH2M HILL, Inc for NACWA and AMWA. 2009

Figure 6-3 Climate change effects on the water sector

Sources: Zack Cunningham/California Department of Water Resources, Oct. 23, 2015 (left); Kelly M. Grow/California Department of Water Resources (right)

Figure 6-4 Aerial view of Lake Oroville, Calif., and the Bidwell Bar Bridge on Oct 23, 2015, showing the water level at 48 percent of historical average or 29 percent of total capacity (left); Lake Oroville's Bar Bridge at 96 percent of total capacity, or 117 percent of historical capacity, on May 11, 2016, in Oroville, Calif. (right)

Extreme drought. Drought events can cause critical water shortages, as has been experienced in multiple water systems in the Southwest. Figures 6-4 and 6-5 show before-and-after photos of California's Oroville Reservoir and of snowpack in the Sierra Nevada mountains, respectively, which supplies much of California's water supply.

Drought risk is projected to persist in the Southwest, as discussed by Cook et al. (2015) and shown in Figure 6-2. The cost of adapting to the drought impacts of climate change is likely to increase, as water providers have to diversify their raw water supply sources. This diversification can mean creating new storage options and interconnections as well as exploring reuse, desalination, and water conservation.

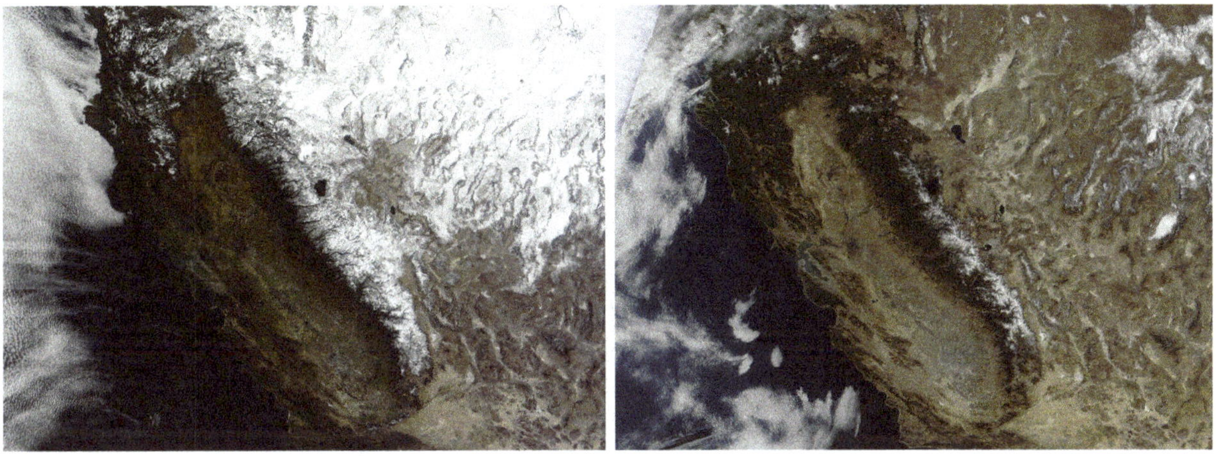

Sources: NASA Earth Observatory

Figure 6-5 Snowpack in Sierra Nevada Mountains, Calif., January 2013 (left) to January 2014 (right)

Heat waves. Extreme heat can lead to a number of operational impacts on water providers, including increased water demand for cooling and irrigation, limitations on work hours, and failure of equipment.

Intensifying storm and flood events. The extreme flood events described in this chapter reflect one of the most likely impacts of climate change: storms of a given recurrence interval are projected to become larger, and storms of a given depth are expected to occur more frequently. This change in rainfall intensity-duration-frequency (IDF) relationships is illustrated in Figure 6-6, which shows projected changes in rainfall IDF by 2100 for Washington, D.C.

Projected changes in rainfall depth and frequency translate to projected changes in flood frequency, which have already been experienced by water providers. For example:

- The Nashville, Tenn., water treatment plant was flooded in 2010, by an extreme storm event larger than any previous event (Figure 6-7).
- The Berwick, Maine, water treatment plant was flooded by two 100-year storms in successive years in 2006 and 2007 (Figure 6-8).
- Loveland, Colo., was devastated by flooding in the Colorado Front Range in September 2013, which caused its water intake dam to breach (Figure 6-9).

Sea level rise. Rising seas will impact coastal water infrastructure in a number of ways. Coastal flooding will be evident at high tides to varying degrees, leading to potential damage to low-lying coastal systems such as treatment plants, pump stations, and power plants. When coupled with more intense tropical storms, sea level rise and storm surge can cause catastrophic damage to water systems, as was experienced in 2012 during Hurricane Sandy and again in 2017 during Hurricane Harvey.

Coastal water supply systems that depend on groundwater or freshwater from upper estuaries will be at risk for saltwater intrusion. Higher ocean levels will drive salt lenses further inland, forcing coastal systems to seek alternate water sources or switch to desalting technologies.

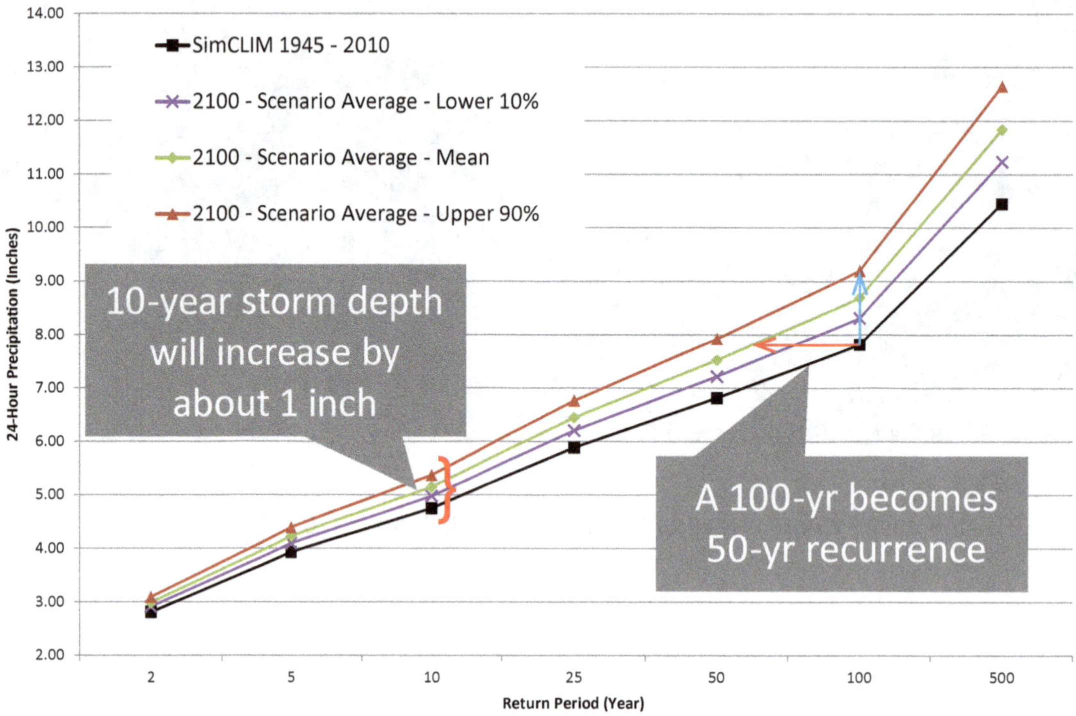

Source: Rainfall Frequency and Global Change Model Options, Technical Memorandum prepared for the City of Alexandria by CH2M HILL (now part of Jacobs), 2011

Figure 6-6 Projected changes in rainfall intensity-duration-frequency relationship for a 24-h storm in Washington, D.C., by the year 2100

Note: SimCLIM is a software tool that is used to facilitate the assessment of risks from climate change.

Source: Photo courtesy of Gresham Smith

Figure 6-7 Nashville Tennessee Water Treatment Plant was shut down in May 2010 due to flooding

Source: Adapted with permission from CH2M Hill, April 8, 2014

Figure 6-8 Berwick Maine Water Treatment Plant was flooded in back to back years (2006 and 2007) by extreme flood events

Source: Tanner Randall, City of Loveland, Colo.

Figure 6-9 Loveland, Colo., water intake dam during floods of September 2013 and showing damage from dam breach after the storm

Costly Impacts of Increasingly Extreme Weather Events on the Water Sector

In an early study of the potential impacts of climate change on the water sector, Association of Metropolitan Water Agencies (AMWA) and the National Association of Clean Water Agencies (NACWA) evaluated the cost of adapting to the range of impacts listed in Figure 6-10 for water and wastewater utilities across the United States through 2050. The costs ranged from $325 million to $692 million for drinking water facilities, above and beyond the costs of asset renewal and replacement or emergency response.

While these potential climate adaptation costs seem daunting, they likely pale in comparison to the costs of no action. Taking no action requires emergency response as well as costly rebuilding due to extreme future storm events, not to mention the indirect economic costs of water service interruption such as business interruption costs, reduced labor productivity, and reduced tourism. While those no-action costs have not been quantified in most climate resilience planning efforts, the available site-specific studies and data indicate that planning for climate change costs less over time than recovering from the damage caused by extreme storm events anticipated under climate change. For example, New York City prepared a wastewater resiliency plan to reduce the risk of damage from sea level rise coupled with future storm surge events similar to what the city experienced during Hurricane Sandy in 2012. Figure 6-11 shows the comparison of cost estimates. Climate change adaptation was estimated to cost $315 million, while the cost of damage from a single storm surge event when coupled with sea level rise in 2050 is estimated to be $1.1 billion, and the costs from recurring events between now and 2050 are estimated to total $2.5 billion. The magnitude of the cost of no action has motivated the city to seek $315 million to fund adaptation. Similarly, Nashville, Tenn., has documented $13 million in costs to recover from the flood of May 2010. Less than half that amount would be required to harden facilities for similar future floods.

Managing the Risk of Climate Change by Developing a Climate Action Plan

All water and wastewater utilities must meet the needs of their customers by balancing service demands with the cost of service. The cost of service reflects a range of risks including asset failure, capacity limitations, and regulatory changes. If shifting climate-induced vulnerabilities are not assessed, water and wastewater facilities will operate with unmanaged risk. Managing this risk requires an understanding of how shifts in climate ranges and extremes will impact each asset over its service life.

A recent Water Research Foundation survey found that the public expects water utilities to plan for climate change. This information is presented in Figure 6-12. Separately,

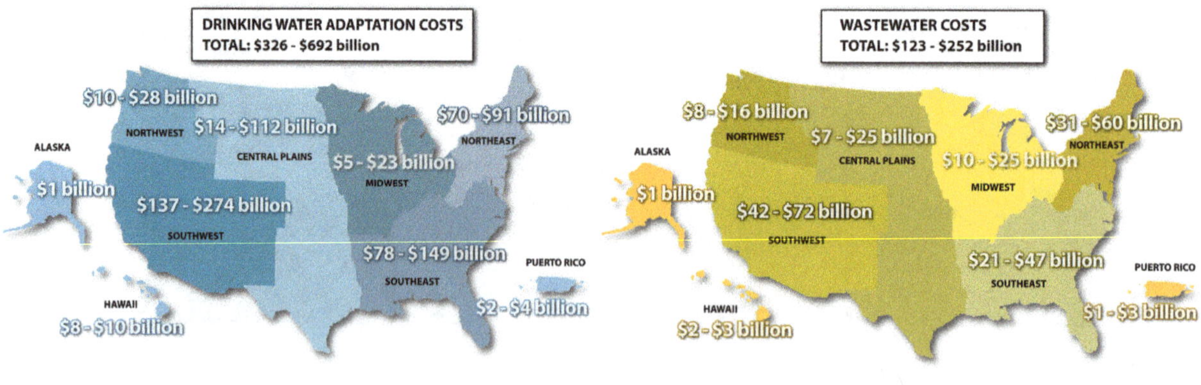

Source: NACWA and AMWA, Confronting Climate Change: An Early Analysis of the Cost of Climate Adaptation, 2009

Figure 6-10 Water and wastewater adaptation costs

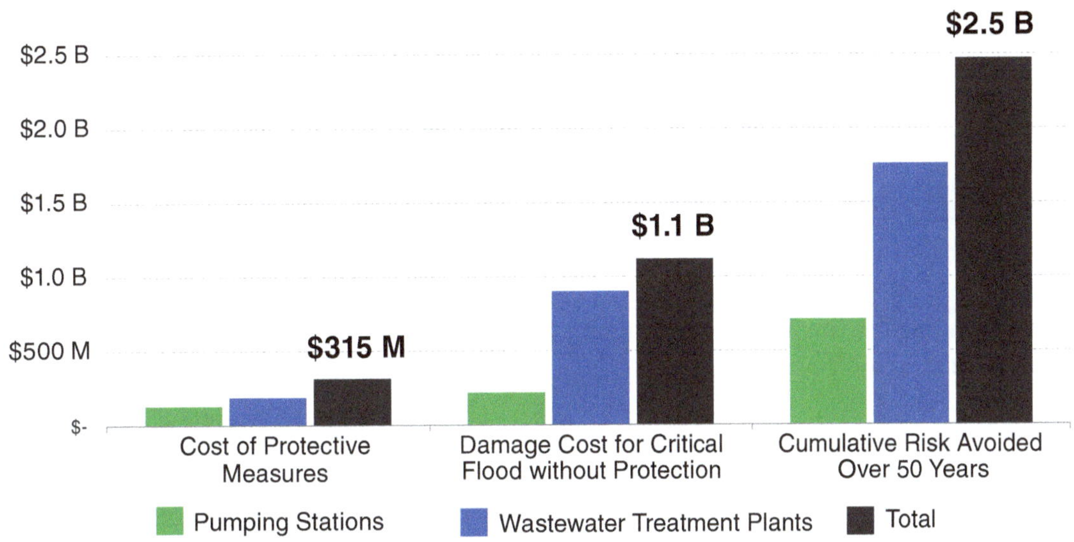

Source: NYC DEP Wastewater Resiliency Plan, 2013

Figure 6-11 New York city comparison of wastewater adaptation costs and cost of no action

the Water Utility Climate Alliance (WUCA) surveyed 13 large water utilities and found that most are considering the threat of climate change in decision processes including immediate operational decisions, asset management planning, and long-term supply planning. These utilities have found that the time for climate adaptation is now. (Katz 2015).

If utilities understand the triggers and thresholds that induce climate vulnerabilities, they can respond proactively to reduce the risk of climate change to water assets. The steps assessing a utility's vulnerability and resilience are outlined in the next section.

VULNERABILITY AND RESILIENCE ASSESSMENT OF A UTILITY

Vulnerability Assessment

Vulnerability to climate change is defined as the degree to which a system is susceptible to the adverse effects of climate change, including climate variability and extremes. Vulnerability is a function of a system's sensitivity and adaptive capacity as well as the character, magnitude, and rate of climate change to which the system is exposed. Therefore, a vulnerability assessment identifies how climate change will uniquely affect a particular water system. Key aspects of a vulnerability assessment include the following.

- Defining the climate variables (heat, drought, rainfall-based flooding, sea level rise and storm surge based flooding, wind, etc.) to which the system is vulnerable, and the magnitude of change anticipated in the planning period. For example, extreme heat can cause failure of electrical systems or impact field staff. Flooding from extreme rainfall or storm surges can damage pumps, motors, and water treatment systems. Extreme wind events can cause power outages that reduce operations or require use of backup power.
- Defining whether the system is exposed to those hazards, such as treatment plants or pump stations in or near the floodplain. Systems near the coast can be flooded by high tides caused by sea level rise.

- Defining how sensitive the system is to exposure. This determination reflects whether the system will be damaged or rendered inoperable if exposed. For example, concrete structures can withstand flooding and will return to service after flooding recedes, whereas electrical and mechanical equipment can be temporarily or permanently damaged by flooding.
- Defining how adaptive the system is to climate impacts. Can the system be readily moved or hardened against damage?

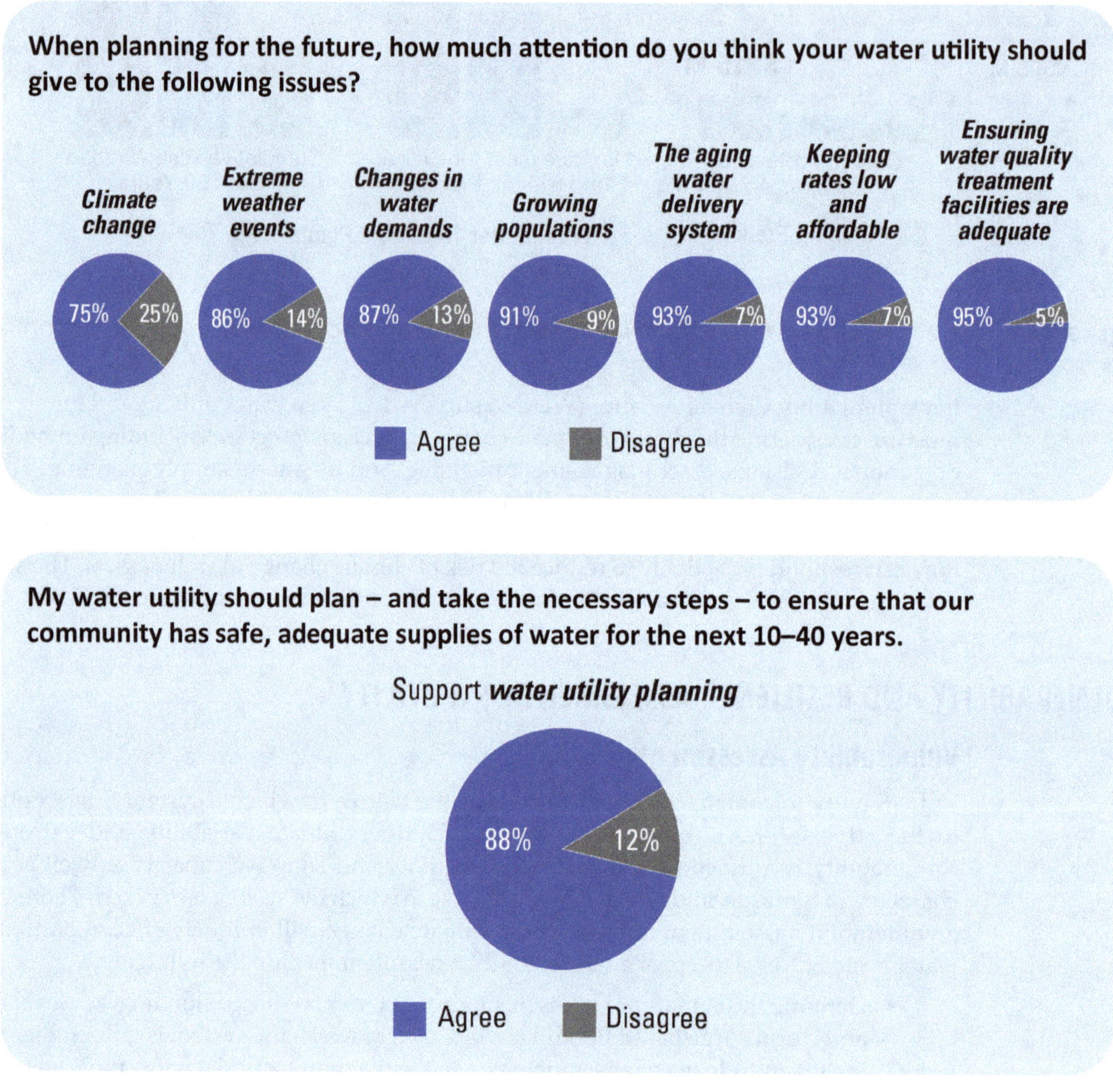

Source: Raucher et al. 2014. Adapted with permission from Water Research Foundation

Figure 6-12 Water research foundation project #4381 survey

Note: Survey found that 92% of Americans want their utilities to take a leadership role in preparing their community for climate change.

Climate Resilience

Climate resilience is characterized by a system's ability to bounce back after a damaging event and its capacity to function despite disturbance. Once climate vulnerability is identified and risks prioritized, vulnerability reduction should focus on increasing the resilience of water systems to future extreme events, as well as planning emergency response actions to address actual impacts.

Climate Adaptation Frameworks

Figures 6-13 through Figures 6-17 provide examples of adaptation frameworks that outline key steps to assess climate vulnerability that can be used to inform the development of a CAP. These frameworks share some common elements:

- Understand climate threats for water systems.
- Define damage thresholds for critical water systems.
- Assess risk.
- Determine adaptation options.
- Implement and monitor.

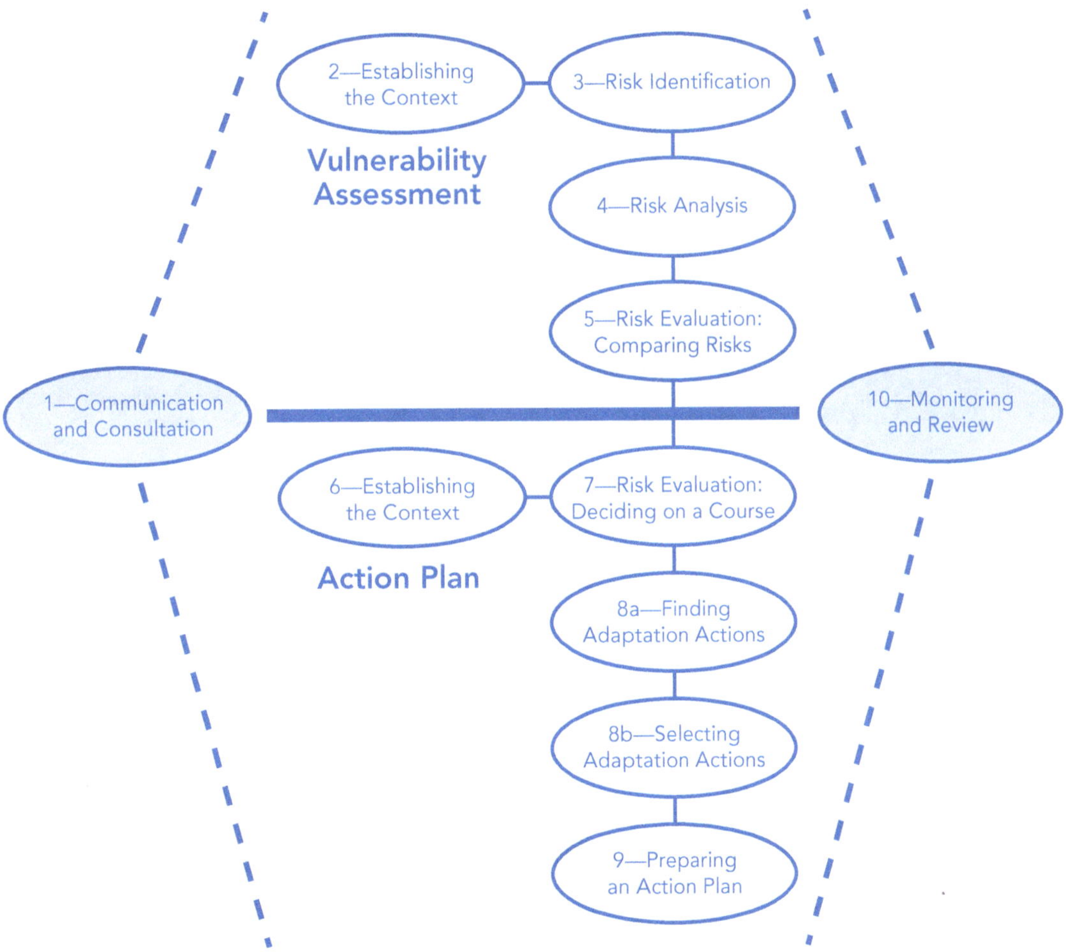

Source: Being Prepared for Climate Change: A Workbook for Developing Risk-Based Adaptation Plans, USEPA, August 2014

Figure 6-13 USEPA's climate ready estuaries roadmap

Note: This figure shows Step 1 through Step 5 of the vulnerability assessment and Step 6 through 10 of preparing an action plan.

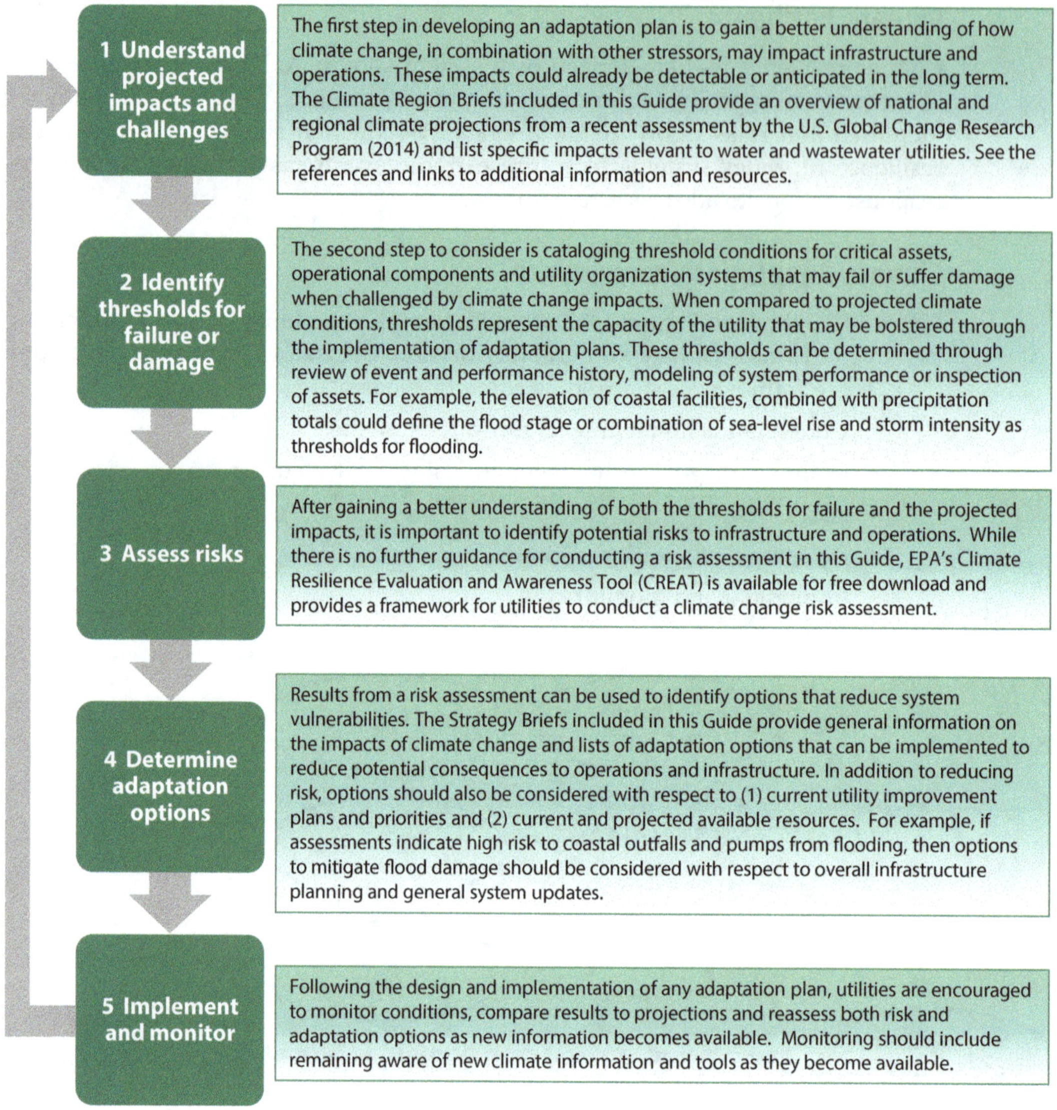

Source: Adaptation Strategies Guide for Water Utilities, USEPA 2012

Figure 6-14 USEPA's climate ready water utilities adaptation planning steps

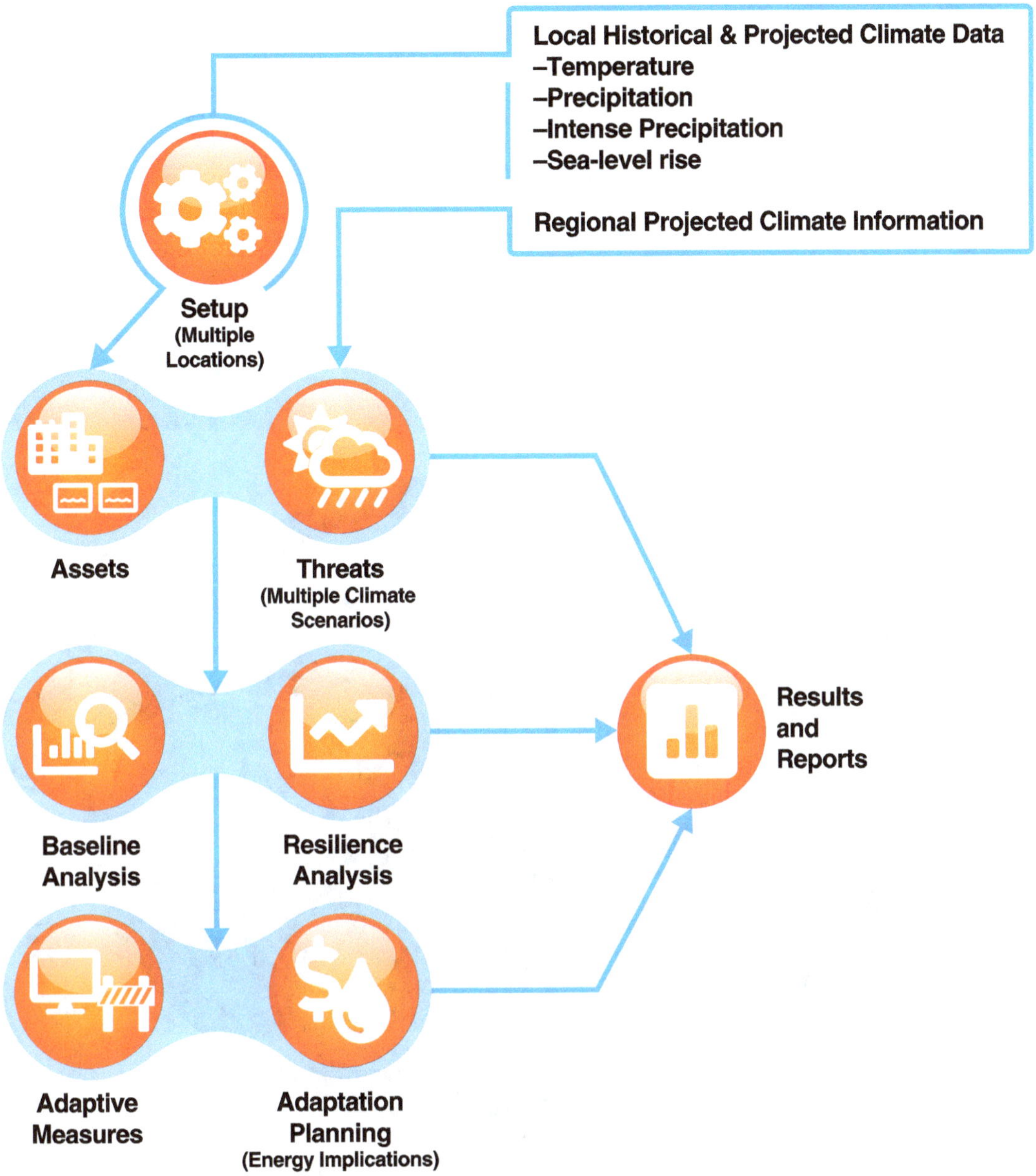

Source: Adapted from: Climate Resilience Evaluation and Awareness Tool, Version 2, Methodology Guide, USEPA, 2012

Figure 6-15 Schematic of process for building climate change awareness to identify threats and potential adaptation plans

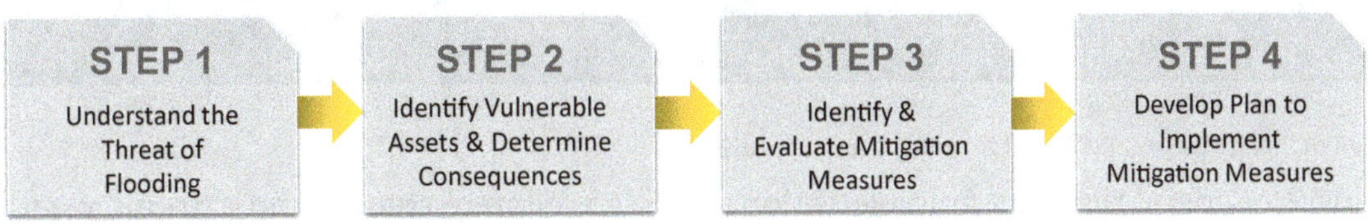

Source: Flood Resilience: A Basic Guide for Water and Wastewater Utilities, USEPA, 2014

Figure 6-16 Flood resilience: a basic guide for water and wastewater utilities

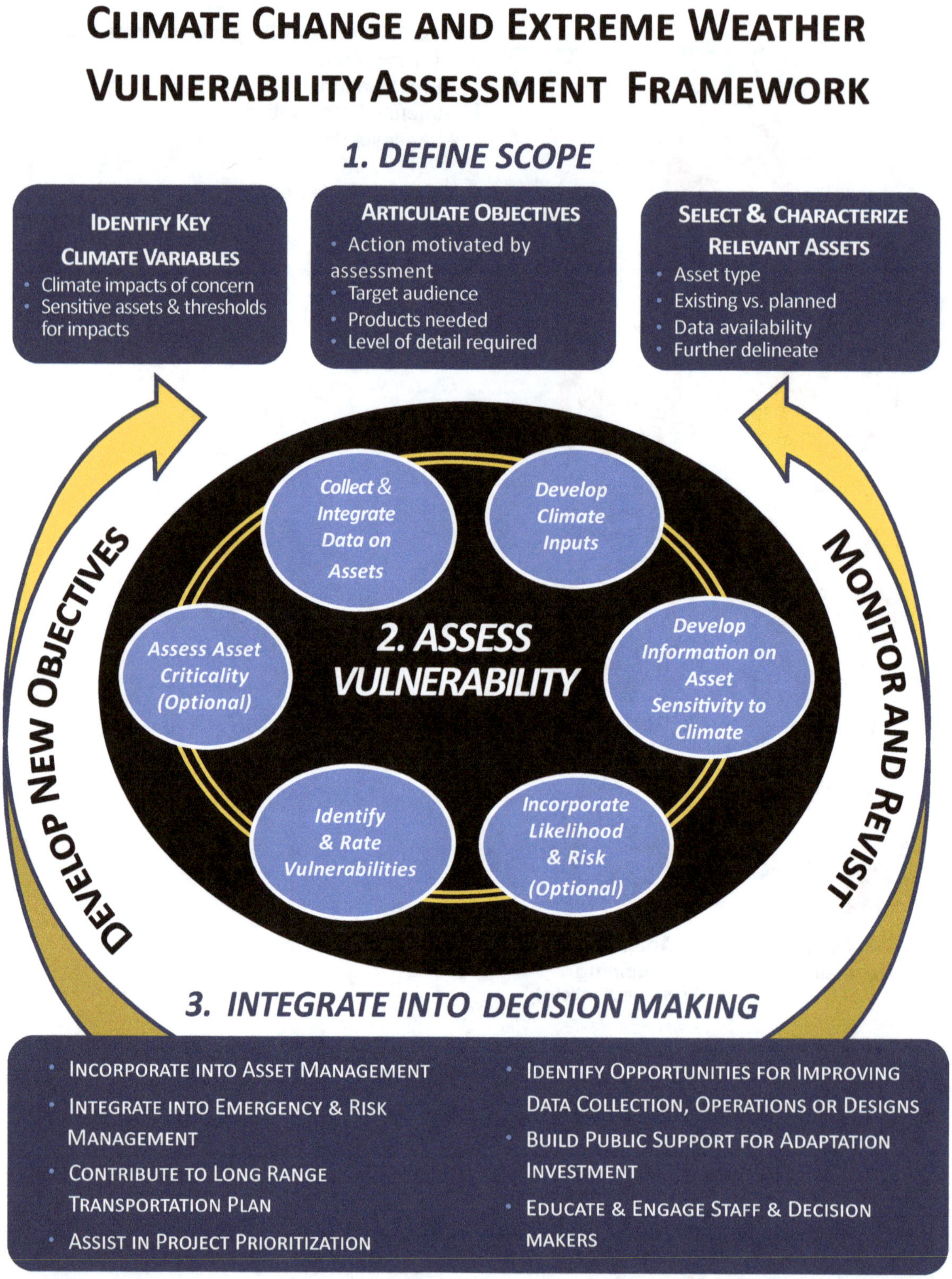

Source: FHWA Publication No: FHWA-HEP-13-005, December 2012

Figure 6-17 Federal Highway Administration's (FHWA's) climate change and extreme weather vulnerability assessment framework

Climate threats are site-specific. Potential threats that could impact a water system include the following:

- More frequent and severe river flooding;
- Coastal flooding due to higher sea level, storm surge, and wave height;
- Increased likelihood of summer droughts, inducing river and reservoir water quality problems and disrupting water supply;
- Coastal erosion and loss of habitat due to higher sea level and increased wave height;
- Saltwater intrusion due to higher sea level, threatening coastal groundwater and estuarine river intakes;
- Increased temperatures and heat waves impacting operations (stress on equipment, higher energy costs, outside worker impacts) and assets (electrical/mechanical systems, pavement areas);
- High-intensity rainfall destroying river habitat;
- Increased wind erosion;
- Declining water quality due to flooding, erosion, and periods of reduced river flow that provide less dilution for discharges;
- Limited water withdrawals due to aquatic life impacts of heat and reduced streamflow;
- Increased treatment costs to provide water supplies;
- Decreased electrical grid reliability due to more frequent and intense storms, impacting reliability of treatment processes and pumping facilities;
- Enhanced plant and algal growth due to increased temperature;
- More runoff, increasing nutrient loading in water;
- Threats to ecological quality of water bodies;
- Increased demand for landscape irrigation and cooling;
- Increased competition for source supplies from agriculture and power;
- Odor impacts.

REFERENCES

Cook, B.I.; Ault, T.R.; and Smerdon, J.E. 2015. Unprecedented 21st-century drought risk in the American Southwest and Central Plains. *Sci. Adv.*, 1, no. 1, e1400082, doi:10.1126/sciadv.1400082.

FHWA (Federal Highway Administration). 2012. Climate Change & Extreme Weather Vulnerability Assessment Framework. https://www.fhwa.dot.gov/environment/sustainability/resilience/publications/vulnerability_assessment_framework/fhwahep13005.pdf. (Accessed: April 3, 2020).

Griffin, D. and Anchukaitis, K.J. 2014. How Unusual is the 2012–2014 California Drought? *Geophysical Research Letters*, 41:24:9017–9023.

Kaatz, L. 2015. Embracing Uncertainty: A Case Study Examination of How Climate Change is Shifting Water Utility Planning. https://urban.csuohio.edu/sites/csuohio.edu.urban/files/Climate_Change_and_Water_Utility_Planning.pdf (Accessed: April 3, 2020.).

NACWA (National Association of Clean Water Agencies) and AMWA (Association of Metropolitan Water Agencies). 2010. Confronting Climate Change: An Early Analysis of the Cost of Climate Adaptation. Englewood, Colo.: CH2M HILL.

NYC DEP (New York City Department of Environmental Protection). 2013. https://www1.nyc.gov/html/dep/html/press_releases/13-107pr.shtml#.XodhMYhKiUk. (Accessed: April 3, 2020).

Raucher, R. et al. 2014. Effective Climate Change Communication for Water Utilities. Water Research Foundation, Denver.

Williams, A.P., Seager, R., Abatzoglou, J.T., Cook, B.I., Smerdon, J.E., and Cook, E.R. 2015. Contribution of anthropogenic warming to California drought during 2012–2014. *Geophysical Research Letters*, 42:16:6819–6828.

USEPA (United States Environmental Protection Agency). 2016. Climate Evaluation and Awareness Tool, v3 Methodology Guide, https://www.epa.gov/sites/production/files/2016-05/documents/creat_3_0_methodology_guide_may_2016.pdf (accessed April 3, 2020).

USEPA. 2014. Being Prepared for Climate Change: A Workbook for Developing Risk-Based Adaption Plans. https://www.epa.gov/cre/being-prepared-climate-change-workbook-developing-risk-based-adaptation-plans (accessed April 3, 2020).

USEPA. 2012. Adaptation Strategies Guide for Water Utilities https://www.epa.gov/sites/production/files/2015-04/documents/updated_adaptation_strategies_guide_for_water_utilities.pdf (accessed April 3, 2020).

Chapter 7

How Do Utilities Prepare a CAP and Keep It Current?

Some climate action plans (CAPs) focus on mitigation only, while others include adaptation. In either case, this chapter discusses the main processes and tools to consider when developing the content of a CAP. In addition to content, developing a CAP requires a number of critical procedural elements. An agency must follow a series of steps that include deciding to develop a CAP, rolling out plan initiatives/goals, communicating goals, and monitoring progress and updating the plan. Critical logistical elements include the following.

i. Convening a climate action planning team
   a) Establishing mechanisms for management support
   b) Including all relevant industry sectors
   c) Seeking support from experts and decision-makers to share information and facilitate a sense of ownership during the development process
ii. Developing stakeholder engagement
   a) Defining community and industry stakeholder groups
   b) Determining the scale of involvement for different stakeholder groups
   c) Defining the role of emergency responders and field staff with detailed site knowledge
   d) Estimating the impacts of adaptation and mitigation on stakeholders and establishing a consultation process to identify and investigate those impacts
iii. Developing a strategy for CAP adoption to sustain public and political support

iv. Developing a communications plan

v. Defining the political players and "downstream" regulatory actions after adoption of the CAP

   a) Defining the balance between policy guidelines and regulations

   b) Identifying the agencies leading a regulatory process that may be needed to achieve the plan goals (if regulations are developed to implement the plan)

   c) Deciding on regulatory enforcement mechanisms (if regulations are developed to implement the plan)

   d) Gathering the legal team to help develop regulations (if regulations are developed to implement the plan)

vi. Establishing the plan monitoring process and update frequency

The procedure to develop, adopt, and implement a CAP must be context-specific. Though this manual cannot offer a "one size fits all" solution, additional guidance is provided in Chapters 8 and 9.

PREPARING A CLIMATE ACTION PLAN: MITIGATION

Though this manual focuses on the impacts of climate change on utility operations, a complete CAP will also include mitigation to address the greenhouse gases (GHGs) produced by the utility. By reducing GHGs, a utility can minimize its contribution to future warming.

GHG emissions from human activities have increased substantially over the last century and are considered the main driver of global warming over the past 50 years (IPCC 2013). Because of past emissions, serious warming of the climate system is now considered inevitable (IPCC 2013, NCA 2014). However, the severity of continued warming, especially beyond 2050, depends on whether global GHG emissions increase, stabilize, or decrease over the coming decades. The scientific community remains uncertain about what concentration of GHGs in the atmosphere will cause catastrophic climate warming, but most estimates suggest that global GHG emissions must fall by 50 to 80 percent by 2050 to avert catastrophic change.

Drinking water and wastewater system operations account for approximately 3 to 4 percent of US energy use and result in more than 45 million tons of GHG emissions annually. For municipal governments, water and wastewater systems are often the largest energy consumers, constituting 35 percent of municipal energy consumption (USEPA 2012).

A climate mitigation plan or GHG emissions reduction plan addresses the energy used by the utility and resulting GHG emissions. It defines the major sources of emissions from a utility's activities, the emission reduction goals, and the approach to achieve those emissions reductions. A GHG emissions reduction plan guides energy and operational investments and should include the following elements:

1. An emissions inventory that documents the energy consumption and GHG emissions of each facility or system operated by the utility
2. An emissions reduction goal stating the level of emissions reduction that the utility will try to achieve by some future date
3. If applicable, a goal assessment documenting that the goal is consistent with local, statewide, or national initiatives
4. A list of GHG emissions reduction strategies that defines how emissions reductions will be achieved
5. An implementation schedule that shows when each strategy will be implemented
6. A monitoring and verification strategy that describes a schedule of actions

GHG Emissions Protocols, Inventory Development, Baselines, and Verification

This section summarizes the major steps, considerations, processes, and tools to inventory GHG emissions. For detailed information, AWWA recommends the Water Research Foundation's "Greenhouse Gas Emission Inventory and Management Strategy Guidelines for Water Utilities" (WRF 2009).

Several protocols exist for documenting GHG emissions. Most protocols are based on the work of the World Resources Institute (WRI) and World Business Council on Sustainable Development (WBCSD), which developed the Greenhouse Gas Protocol, setting the global standard for how to measure, manage, and report GHG emissions. Additional GHG accounting protocols have been developed for specific sectors, purposes, and organization types based on the framework established by WRI and WBCSD. Though no specific protocol has been developed for water utilities, several existing protocols provide appropriate frameworks that water utilities can use. A few relevant protocols are described in Table 7-1. Each protocol establishes rules for developing an emissions inventory. While each protocol is unique, all share several common elements. All protocols require:

- Assessment of all six GHGs regulated under the Kyoto Protocol: carbon dioxide, methane, nitrous oxide, hydrofluorocarbons, perfluorocarbons, and sulfur hexafluoride
- Establishment of an "organizational" or "system" boundary that defines what is included in the inventory
- Reporting of Scope 1 and Scope 2 emissions and optional reporting of Scope 3 emissions

Emissions types. Following are descriptions of three difference emissions types: Scope 1, Scope 2, and Scope 3.

- Scope 1 emissions are those emitted directly from sources onsite. This type of emission includes emissions from treatment processes, tailpipe emissions from fleet vehicles, and emissions from onsite stationary combustion.
- Scope 2 emissions are those associated with consuming purchased electricity, steam, heating, and cooling.
- Scope 3 emissions comprise all other indirect emissions. They may come from processes outside of the organizational or system boundary of the entity. Scope 3 emissions may include emissions from water extraction and conveyance processes embedded in wholesale water purchases or emissions from a contractor's disposal of residual water treatment solids. Reporting Scope 3 emissions is optional, but this category often represents important emission sources. Additionally, future emission reduction strategies or changes in operations can often change the classification of emissions between scopes.

Data management. Creating a GHG inventory requires gathering data on electricity usage, fuel usage, and operations. A well-designed data management system will work with the utility's existing operational structure and organization, account for data at the right level of detail, and allow the utility to query the data in useful ways.

Detailed information on the design of data management systems is beyond the scope of this manual. However, utility managers may consult several excellent resources for assistance. Additionally, Appendix A of The Climate Registry's "General Reporting Protocol Version 2.0: Managing Inventory Quality" provides information about establishing data acquisition and management systems as well as standards to ensure that these systems meet quality requirements. The Climate Registry staff also help members develop customized data management systems for emissions reporting.

Table 7-1 Useful greenhouse gas protocols

Protocol	Description	Source
GHG Protocol	The global standard for how to measure, manage, and report GHG emissions	GHG Protocol
US Public Sector Protocol	Based on the Corporate Standard framework but specifically configured for the unique structures and needs of US government operations at the federal, state, and local level	GHG Protocol
GHG Protocol for Cities (jointly with ICLEI*)	Specifically designed for use by cities and other local level government structures that provide public services	ICLEI USA GHG Protocol
Project Protocol	Comprehensive, policy-neutral accounting tool for quantifying the GHG benefits of climate change mitigation projects	GHG Protocol
The Climate Registry	Operates the North American voluntary GHG registry where organizations can publicly report their GHG emissions and document progress toward achieving emissions reductions	The Climate Registry
General Reporting Protocol (GRP)	A general reporting protocol that can be used by a wide array of businesses and public agencies	The Climate Registry
Electricity Power Sector (EPS) Protocol	Provides important supplemental reporting guidance, additional requirements for the EPS, and sector-specific clarifications. Designed as an appendix to the GRP	The Climate Registry
Local Government Operation Protocol (jointly with ICLEI)	A program-neutral protocol that is designed to allow local governments to quantify and report GHG emissions resulting from their operations	The Climate Registry
Water-Energy-GHG Protocol (In process/Not yet available)	A water sector–specific protocol to allow water agencies to document the energy and GHG emissions embedded in various water products including potable, recycled, and treated wastewater	The Climate Registry
ICLEI - Local Governments for Sustainability	A movement of local governments pursuing deep reductions in carbon pollution and tangible improvements in sustainability and resilience.	ICLEI USA
GHG Protocol for Cities (jointly with GHG Protocol)	See above	ICLEI USA GHG Protocol
Local Government Operation Protocol (jointly with TCR)	See above	The Climate Registry

* ICLEI was formerly known as International Council for Local Environmental Initiatives, but now just goes by its initials and the tagline "Local Governments for Sustainability."

Emissions reporting verification. The final step to emissions reporting is third-party verification. This step is optional but will improve the quality and reliability of the emissions report. Verification ensures that GHG inventories are accurate and complete. Most protocols have unique verification requirements and may provide a verifier certification. The Climate Registry has a General Verification Protocol that documents the entire verification process, from choosing a certified verifier to submitting verified emissions reports.

Projecting future emissions. Once a utility has an inventory of current emissions, a next step can be to project how emissions may change in the future. This step involves making assumptions about future conditions such as

- population,
- water demand,
- climate,
- supply availability,
- regulations, and
- energy sources and emissions factors.

Projections should be consistent with assumptions made for other utility planning processes. In most cases, projected emissions can be based on recent/previous planning assumptions to estimate the expected energy demand and operations related to water extraction, conveyance, treatment, and distribution processes.

In these calculations, utilities should consider how energy source emissions factors are likely to change in the future. Thirty-seven states now have renewable energy portfolio targets (voluntary) or standards (mandatory). Over the coming decades, the portfolio of energy resources in these states is likely to shift toward renewable energy generation, reducing the average emissions associated with purchased electricity. Utilities that purchase energy from a specific energy utility should consider the long-range energy portfolio plan for that specific utility and work with the utility to identify how emissions factors may change in the future. Smaller sources of emissions (such as of fleet vehicles and pumps) may also vary over time as the efficiency changes.

Establishing a GHG Emissions Reduction Goal

Once a utility has documented current emissions and projected future emissions, the next step is to establish a GHG emissions reduction goal. An emissions reduction goal has many benefits, including

- motivating climate action throughout the agency;
- showing management support and expectations for climate leadership efforts;
- demonstrating environmental commitment to the community;
- identifying the agency as a leader in the water sector;
- motivating identification of new GHG emission reduction opportunities;
- complying with regulation;
- taking advantage of a regulatory streamlining process;
- encouraging operational innovation;
- reducing operational energy and costs;
- improving employee morale, recruiting, and retention; and
- drawing positive attention from stakeholders.

Baselines. Before a utility can establish a reduction goal, it must determine its baseline level of emissions. To determine the level from which GHG reductions will be measured, it is recommended that at least three years of past emissions data be considered to understand the level of emissions emitted each year, the major sources of emissions, and how annual emissions fluctuate. The baseline can be set as the level of emissions from a single year within the past three years, or as an average of emissions over a period of years. If a single year of emissions is used as the baseline, the most recent year of available emissions data is preferred.

For water and wastewater utilities, interannual variability in weather or other operational considerations may result in substantial swings between annual emissions rates. Hot and dry years may result in greater demand for irrigation water and therefore higher energy use and emissions for the utility. Wet years may result in higher hydroelectric generation that meets more of the utility's energy demand without emissions. Utilities should carefully consider the ramifications of interannual variability on their operations when establishing both baselines and reduction goals.

If a utility's annual emissions vary significantly, a multiyear average baseline will smooth those fluctuations. However, some GHG evaluation programs favor single-year emissions baselines, and selecting a multi-year average baseline may hinder public recognition for goal setting. For example, neither the US Environmental Protection Agency (USEPA–co-sponsored Climate Leadership Awards program nor The Climate Registry's Climate Certified program recognizes multiyear averages for baselines.

If both public recognition and interannual variability are concerns for the utility, one solution is to strategically select emissions data of a single year that most closely approximates average conditions or that is least affected by anomalous conditions. Water utilities may also consider choosing a baseline year that aligns with a state, county, or city baseline to show consistency with overarching emissions reduction plans.

Reduction goals. The utility should scrutinize the GHG emissions inventory to identify the major sources of emissions. Each utility has a unique set of GHG emissions sources and reduction opportunities, so goal setting is highly customized. However, there are a number of standard considerations.

- The goal should cover emissions from all of the utility's activities.
- Emissions goals can be expressed as percent or absolute GHG reductions or GHG-specific intensity reductions (e.g., GHG/mgd).
- Reduction goals should be achieved in the near future (5–15 years) and may also include additional longer-range goals (20–50 years).
- Goals should be at least as aggressive as state and local emission reduction targets. Goals should be significantly more aggressive than business-as-usual expectations, even if emissions are expected to decrease.

GHG emissions reduction goals should be challenging and designed to drive innovation and investment. For a reduction goal to demonstrate the utility's environmental commitment and establish the utility as an industry leader, the goal must be aggressive. USEPA has developed a Corporate GHG Goal Evaluation Model (2014) to evaluate the extent to which reduction goals represent actual GHG reductions as opposed to industry-wide expected business-as-usual changes.

As for baselines, interannual variability poses a challenge for establishing reduction goals. Variability can cause a utility to fail to achieve its goal because of anomalous weather or operational conditions in the goal year. Some agencies have used a multiyear trailing average to gauge whether the target has been achieved while smoothing out annual variations. Unfortunately, a multiyear average reduction goal will mask the effect of GHG emissions reduction actions implemented during the later years of the averaging period.

GHG reduction goals must define a magnitude of reduction and a time frame within which the reduction will occur. Utilities will have to consider what is possible within their operational and financial constraints. In some cases, a goal is established by a board or director who wants to achieve a specific level of emissions reduction. In other cases, a goal may be established after several GHG emissions reduction strategies have been evaluated. Either strategy can be effective. In both cases, the goal's time frame must balance the time required to plan, approve, fund, and implement GHG reduction projects with the need to show that actions are being taken.

Some utilities establish a near-term goal within 5 to 15 years that is based on the "feasible pathways" approach described above, and a long-term goal of 20 to 50 years that is meant to drive innovation. The California Department of Water Resources (CDWR) won a Climate Leadership Award for Excellence in GHG Management in 2015 for their GHG emissions reduction plan. The plan included two goals:

- 33 percent reduction from 2010 emissions levels by 2020
- 77 percent reduction from 2010 emissions levels by 2050

When CDWR established the goals, it had clear plans for several GHG emissions reduction strategies that it could implement to achieve the 2020 goal, but it will have to devise new strategies over the coming years to achieve its 2050 goal.

Strategies for Reducing GHG Emissions

Once a reduction goal has been established, the next step is to identify GHG emissions reduction strategies. Potential strategies for reducing emissions are likely to fall into several categories:

- Improve energy efficiency
- Manage and develop renewable energy supplies
- Self-generate electricity
- Change business operations and equipment use
- Improve water efficiency

Identifying feasible emissions reduction strategies often requires a team approach, engaging diverse utility staff, outside consultants, and other stakeholders. Chapter 8 of the manual discusses effective team assembly, stakeholder involvement, and decision-making. Several resources can assist water utilities with developing GHG emissions reduction strategies, including the following:

- WRF Energy Management Knowledge Portal
- Greenhouse Gas Mitigation Planning: A Guide for Small Municipal Utilities
- Greenhouse Gas Emission Inventory and Management Strategy Guidelines for Water Utilities
- Energy Efficiency in Water and Wastewater Facilities: A guide to developing and Implementing GHG Reduction Programs

Once a list of potential GHG emissions reduction strategies has been developed, the team must filter the list to identify the best options. Initial screening criteria suggested by Fernandez et al. (2005) include the following.

1. Implementation feasibility: commercial implementation of the option by the utility must be currently feasible.
2. Uncertainty: the option must be relatively certain to be effective, with a low possibility of negative effects to human health or the natural environment relative to the status quo.
3. Regulatory restrictions: the option must not conflict with existing regulatory policies or frameworks.

Individual utilities may add additional criteria specific to their constraints or interests. For example, a utility may discard any option that doesn't produce co-benefits such as air quality and water quality improvements within its service area. If timing is critical, any option that cannot be implemented in the short-term may be ruled out.

Once the list of options has been screened, the utility conducts a detailed analysis of each option to quantify the expected emissions reductions and the cost per unit of reduction associated with each strategy. A number of tools are available to help utilities explore GHG and energy efficiency strategies. WESim is an easy-to-use analytical tool that can be used to evaluate the energy and GHG implications of water management decisions. The Excel-based model allows the user to explore a range of possible scenarios, such as increased demand for water resources, development of alternative water and energy sources, and needed water treatment improvements resulting from emerging contaminants and changing water quality guidelines (Pacific Institute 2012).

Other tools for exploring GHG and energy efficiency strategies include

- WaterSmart Tool Kit,
- Water-Energy Sustainability Tool, and
- Water Conservation Tracking Tool.

Quantitative tools like these may not capture all of the benefits of some mitigation strategies. For example, an upgrade to a utility's water treatment plant may yield both GHG benefits and water quality benefits, or other benefits that are difficult to monetize, such as improved water taste. It is a good idea to track these ancillary benefits as they may affect decisions later in the process or funding eligibility.

Next, the potential GHG mitigation strategies should be compared. Table 7-2 provides a simple example of a strategy comparison worksheet. However, strategy selection is rarely as simple as choosing the strategy that yields the most GHG savings per dollar spent. Many other issues affect decisions, including public perception, future regulatory expectations, and local investment versus investment that sends money out of the community. Strategies that yield some GHG benefit, even at high marginal costs per metric ton, while helping the utility achieve other goals often win over strategies that yield only GHG benefits.

Some strategies, such as renewable energy purchases or development, are scalable. Emissions savings will depend on the level of development. Scalable strategies are often attractive because a utility can implement them at a low level and then scale up if the strategy is successful.

Deciding which strategies to include in the GHG emissions reduction plan should involve high-level decision makers as well as the group that developed and analyzed the strategy list. Implementing these strategies requires allocating funding, so people who have the authority to authorize such expenditures must be involved in the decision.

This group must decide which strategies go forward, at what level each strategy is implemented, and a timeline for implementation. Strategies must be compared to the utility's projected future emissions to ensure that the portfolio of strategies will meet the targeted emissions reductions. Utilities should plan for at least 10 percent more emissions reductions than necessary, to hedge against strategies that do not produce the expected level of reductions or take longer to implement than anticipated.

Implementation schedules. GHG emissions reduction strategies are typically implemented over time, often phased in as system components are scheduled for upgrades. These timelines should be described in the plan, identifying when each strategy will be implemented and any dependencies or contingent projects. While some GHG reduction strategies will be implemented as standalone projects, others will require changes to a utility's operating procedures. In these cases, it is important to lay out how the strategy will be integrated into the operational decision-making process. For strategies such as peak shaving that involve discretionary operations decisions, the plan must define when and how decisions will be made about modifying operations.

Table 7-2 Example GHG Reduction Strategy Matrix

Measure Information					Cost Analysis			Co-Benefit Qualitative Analysis			Implementation Analysis		
Mitigation Measure Name	Evaluation Size	Potential Capacity	Assumed Lifetime	Annual GHG Reductions (MT)	Initial Capital Cost	Marginal Cost per Metric Ton	Total Cost	Description of Co-Benefit	Co-benefit value	Quantified Co-benefit value	Technical Feasibility	Implementation Immediacy	Mechanism In Place?
Renewable Energy													
Renewable Energy Purchase Agreement	10 GWh/ year	unlimited	20	3,000		$250	$750,000	NA	NA		High	High	Green Energy
Onsite (owned) Renewable Generation	100 MWh/ year	2 GWh/ year	20	30	$300,000	$500	$300,000	Public perception, energy price stability	low-medium		Medium	Medium	Yes
Energy Efficiency													
Refurbish Pump #1&5	2 pumps	12 pumps	15	20	$250,000	$833	$250,000	Reduce overall energy demand	medium		Medium	Medium	Yes
Water Treatment Plan Upgrade	1 plant	3 plants	30	100	$15,000,000	$124,500*	$15,000,000	Water quality improvement, taste improvement, energy savings, **meets potential new regulatory requirement**	high	85,000/yr			
Install Scada System	2 zones	5 zones	30	15	$1,000,000	$222	$1,000,000	Reduce overall energy demand, Reduce labor costs	High	$30,000/yr	Medium	Medium	Yes
Building Energy Retrofits (HVAC and lighting)	All offices	All offices	20	20	$200,000	$0 cost savings cover costs	$200,000	Annual energy savings, improved workspace conditions	High	10,000/yr	High	High	Yes

Continued

Table 7-2 Example GHG Reduction Strategy Matrix (*Continued*)

Measure Information					Cost Analysis			Co-Benefit Qualitative Analysis			Implementation Analysis		
Mitigation Measure Name	Evaluation Size	Potential Capacity	Assumed Lifetime	Annual GHG Reductions (MT)	Initial Capital Cost	Marginal Cost per Metric Ton	Total Cost	Description of Co-Benefit	Co-benefit value	Quantified Co-benefit value	Technical Feasibility	Implementation Immediacy	Mechanism In Place?
Offsets and Credits and RECs													
United States Offset Programs	20,000 MT CO2e/yr	unlimited	NA	20,000		$5 to $15	$100T to $300T	NA	NA		High	High	Carbon Market
International Offset Programs	20,000 MT CO2e/yr	unlimited	NA	20,000		$5 to $10	$100T to $300T	NA	NA		High	High	Carbon Market

* May be required in the future

Monitoring and verification plan. As for any long-term project, periodic evaluations of GHG emissions reduction progress are essential. Submitting annual emissions reports to an emissions registry ensures that the utility has an annual data point showing its incremental progress toward an end goal. However, data from each individual year can be noisy and may not always show a clear signal of progress. In addition to annual monitoring, detailed progress reviews at three- to five-year intervals are also recommended. At these in-depth reviews, performance of each emission reduction strategy should be considered, and the implementation schedule should be modified if needed.

While no state directly regulates GHG emissions from water utilities, there are a few regulatory programs that offer a streamlined review process for projects that are consistent with a programmatic emissions reduction plan. If a utility intends to use its GHG reduction plan to take advantage of a streamlined regulatory review, monitoring and verification will be important compliance mechanisms. Each regulatory program is different, but one example is the California Environmental Quality Act (CEQA), which allows individual projects to tier off of a programmatic plan for the reduction of GHGs in order to streamline the cumulative impacts analysis of specific projects. CEQA guidelines section 15183.5 requires that an eligible plan "establish a mechanism to monitor the plan's progress toward achieving the level [of GHG reduction] and to require amendment if the plan is not achieving specified levels [of GHG reduction]"(CEQA, 2008).

Example: Scope 3 Emissions Accounting

Consider two hypothetical utilities, A and B. Both Utility A and Utility B receive about 20 percent of their water supplies from a wholesale water supplier. They both have their own local water supplies that make up the balance of their water supply portfolio. Utility A has included the Scope 3 emissions from the extraction and conveyance of wholesale water supplies in its emissions inventory, while Utility B has not. Both capture the emissions from extraction, conveyance, treatment, and distribution of local supplies in their Scope 1 and Scope 2 emissions. After several years of emissions reporting and progress toward their emissions reduction goals, the boards of both utilities decide to prioritize local water supplies and minimize the amount of water being purchased from the wholesaler. After working to increase local supplies, both utilities cut their wholesale water purchases to 10 percent of supply. Utility A finds that its Scope 3 emissions have been cut in half and its Scope 1 and 2 emissions have increased by about 15 percent, resulting in overall emissions decreasing by about 5 percent. Utility B has never included Scope 3 emissions in its GHG considerations; thus, it only sees the 15 percent increase in Scope 1 and Scope 2 emissions. Both utilities have actually reduced GHG emissions, but only Utility A, with its comprehensive accounting of emissions, fully captures the effect of the change in water supply.

PREPARING A CLIMATE ACTION PLAN: ADAPTATION

This section summarizes the major elements of a CAP with a focus on adaptation, as well as references and tools that can help utilities develop the information in each step. For more detailed information, AWWA recommends consulting the references and resources cited under each CAP component described below.

More resources are available from other agencies. Chapter 10 describes case studies that will help identify the scope of other CAP efforts and the process and tools used by other agencies in developing content and conducting a stakeholder process. USEPA's Creating Resilient Water Utilities initiative (https://www.epa.gov/crwu/) describes processes, tools, and other resources. Examples of climate change vulnerability assessments and adaptation efforts across the country can be found in the "Case Study and Information Exchange" page of USEPA's resilient water utilities page. This geodatabase provides examples of how agencies are working to assess vulnerability and adapt to impacts on drought, floods, ecosystem, service reliability, and water quality. This platform can be accessed at https://epa.maps.arcgis.com/apps/MapSeries/index.html?appid=03d35ca84b5944f8b3ab59bf3a981462.

Many of the agencies that have listed project examples in USEPA's Case Study and Information Exchange used the CREAT tool to develop climate change adaptation plans. CREAT uses utility information, such as population served, total flows, and financial conditions, to help agencies assess risk portfolios. Major threat categories within the tool are drought, ecosystem changes, floods, service demand and use, and water quality degradation. Case Study agencies used these identified risks and threats to develop adaptation measures. The CREAT can be accessed here: https://www.epa.gov/crwu/creat-risk-assessment-application-water-utilities.

In addition to USEPA, the Water Utility Climate Alliance (https://www.wucaonline.org/) has produced a number of guidance documents that can help utilities develop a CAP. The following sections refer to publications with particular relevance for defining vulnerabilities and evaluating adaptation measures.

Components of a CAP With an Adaptation Focus

Chapter 5 described the main components of a CAP as it relates to climate change adaptation, and Chapter 6 presented general frameworks for developing vulnerability analyses and CAPs (See Figures 6-13 to 6-17). Based on those frameworks, a general process can be defined, focused on assessing vulnerabilities and evaluating the adaptation measures that will become key recommendations of the CAP. Figure 7-1 presents the relationship between CAP adaptation components.

The CAP for a given utility will be developed, adopted, and implemented in the context of larger policies, plans, and regulations—federal, state, regional, or local—that may be related to climate change or infrastructure resilience. It is important to review existing policies, plans, and regulations not only to align goals but also to achieve synergy. Aligning with existing documents can yield access to expertise, stakeholders, data, and implementation resources used in the development of other policies and plans. Additionally, cross-checking actions recommended with potential CAP adaptation measures avoids redundancy and helps identify elements that can be leveraged to maximize the value of specific recommendations of the CAP.

Vulnerability assessments begin by evaluating the exposure and sensitivity of local geography, assets, customers, and operations to specific hydroclimatic variables, as described in Chapter 6. The utility's adaptive capacity can be assessed in that context. Many sectors are impacted, such as water supply, water demand, hydropower, wastewater collection, wastewater treatment, and habitat. To identify sectors to be explored in defining exposure and sensitivity, consider the vulnerability matrix developed by California Department of Water Resources (USEPA Region 9 CADWR 2011). This concept of multiple-sector vulnerability is an array for assessing exposure, sensitivity, and adaptive capacity.

Because resources are limited, vulnerabilities should be prioritized. Prioritization will dictate the emphasis and timing of adaptation measures. However, prioritizing vulnerabilities can be politically challenging, given that stakeholders come to the process with different priorities and values. This step needs to be conducted with input from key stakeholders and managed by professionals who can help a group achieve consensus. Multiple resources describe methodologies for conducting a stakeholder-driven prioritization process, including the Delphi method (RAND 1969), participatory decision-making (USBR 2012), and value-focused processes based on multi-criteria ranking.

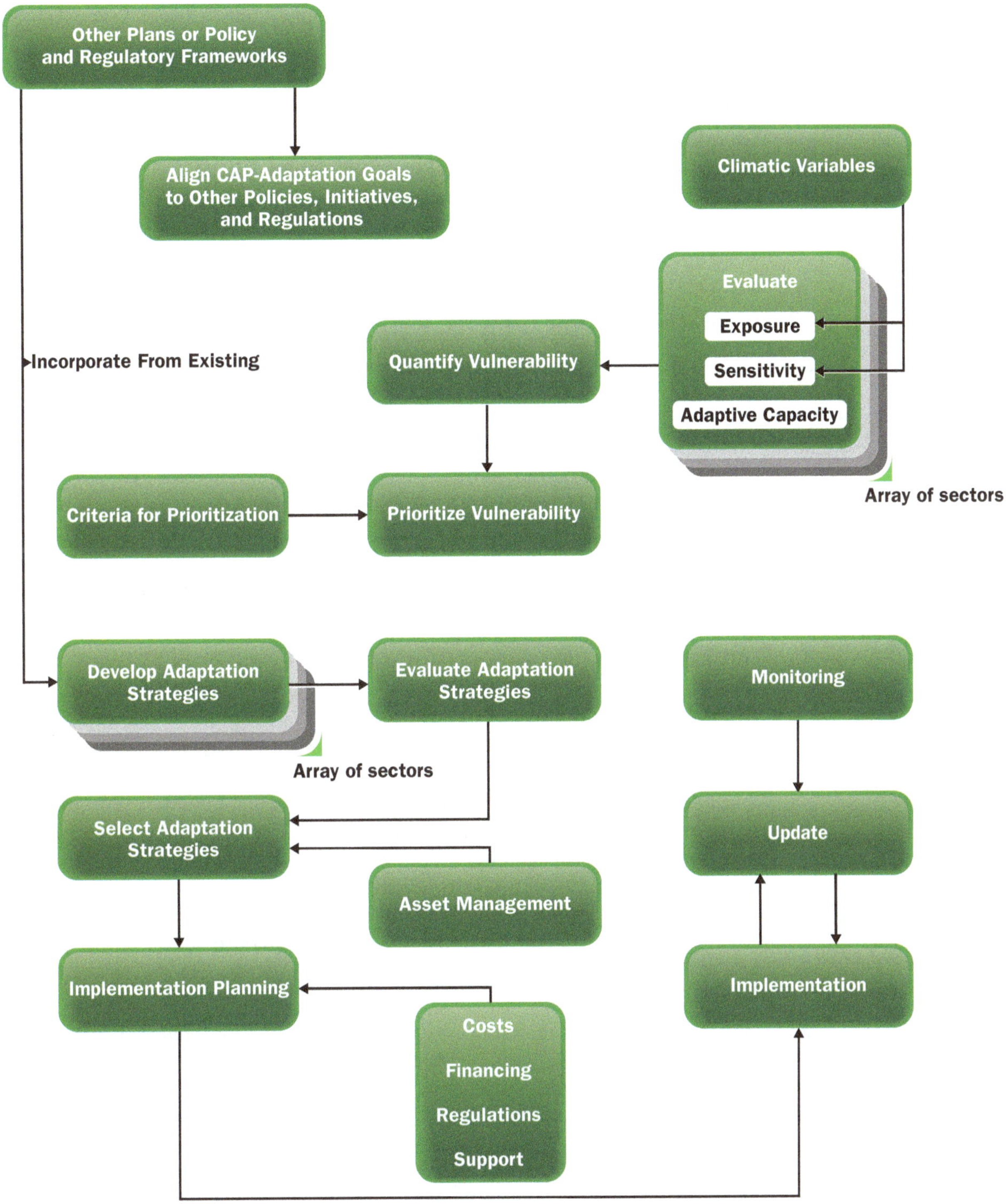

Figure 7-1 Main steps in the development process for an adaptation CAP

The processes for defining adaptation strategies and implementing CAPs are described in this chapter. It has been shown that adaptation strategies need to be designed based on priority vulnerabilities. Like exposures and sensitivities, adaptation strategies are often multi-sectorial.

To evaluate specific adaptation strategies, an achievable planning goal must be defined. The planning goal informs not only the prioritization of vulnerabilities but also the evaluation of adaptation strategies and, ultimately, the scope of the CAP. Planning goals are informed by key decision-makers and stakeholders. The following section describes considerations for defining planning goals.

Assessing the Context of the Climate Action Planning Effort

Chapter 6 described reasons to develop a CAP. Each utility has a unique context, and not all reasons will be applicable to all agencies. Deliberate assessment of the utility's stakeholder, political, policy, and regulatory context is necessary to establish realistic planning goals.

Two initial questions for setting planning goals are as follows.

1. Do state, county, or local government entities have policies that support or encourage climate action planning?
2. Are customers supportive and engaged?

When considering the level of public support for a CAP, utilities should consider the information presented in Figure 6-12, summarizing results from a Water Research Foundation Project (Project #4381) showing that the public expects their utility to take a leadership role in preparing their community for climate change (Raucher et al. 2014). Reviewing the survey information can help the utility prepare a business case for the development and implementation of a CAP. In launching the planning effort, utilities should consult with public engagement and public relations experts to guide the utility's communications.

After assessing the level of stakeholder engagement for a CAP, the utility must gauge its own capacity for the planning effort. Considerations include the following.

1. What tools does the agency already have in place for planning? Resource analysis? Demand analysis? Hazard planning?
2. Have climate vulnerabilities been a focus of past planning efforts? Have recent events such as drought, storms, or extreme temperatures revealed vulnerabilities that need to be addressed?
3. What information (data, models, tools) is already available within the agency or from federal, state, and local governments; academia; foundations; and others that would help inform climate action planning? Useful information and tools are presented in the multiple case studies listed in Chapter 10 of this manual, as well as in USEPA's "Adaptation Case Study and Information Exchange." Additionally, the 2015 WUCA document "Actionable Science in Practice" describes the "chain of models" commonly applicable for vulnerability assessment and evaluation of adaptation measures (WUCA 2015).

The planning effort must be laid out with a specific idea of what can realistically be achieved given the utility's resources. Multiple levels of analysis can be conducted during CAP development, from highly analytical and sophisticated to more qualitative. Sophistication applies in two dimensions, as discussed in CDWRs Climate Change Handbook for Regional Water Planning (USEPA Region 9 CADWR 2011). The first dimension is the forecast of hydroclimatic variables, and the second dimension corresponds to the quantification of future climate impacts on the relevant sectors of the water resources system and on sea level. This concept is illustrated in Figure 7-2.

Assessing Vulnerabilities and Impacts

In order to assess vulnerabilities and the impacts to the system, first it is important to understand what vulnerability is. One framework to understand vulnerability defines it as a composite of three main components:

Vulnerability is the degree to which a system is susceptible to, or unable to cope with, adverse effects of climate change, including climate variability and extremes (IPCC 2007).

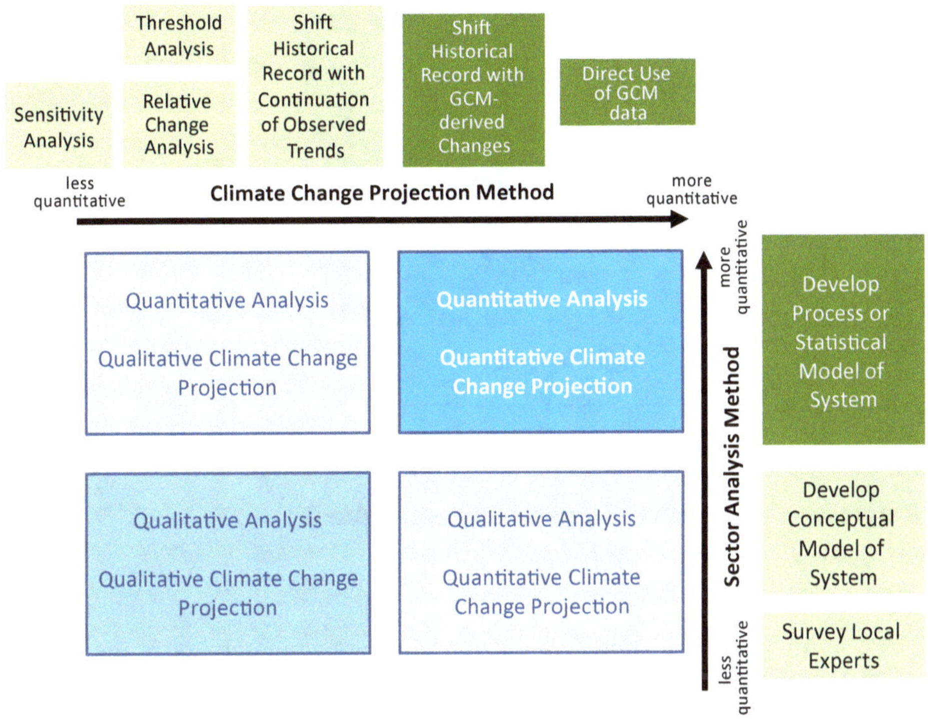

Source: Climate Change Handbook for Regional Water Planning, USEPA Region 9, CADWR, 2011

Figure 7-2 Quantitative and qualitative climate change analysis

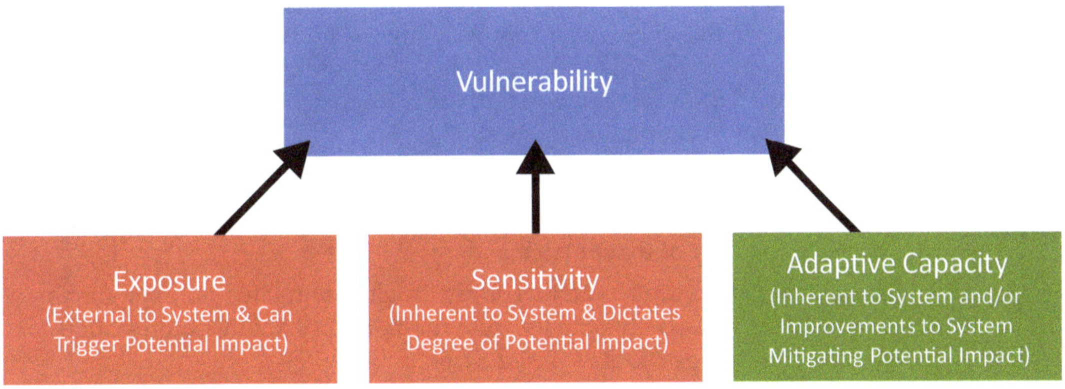

Figure 7-3 Vulnerability as a function of exposure, sensitivity, and adaptive capacity

Vulnerability is a function of the character, magnitude and rate of climate change, and the degree to which a system is exposed, along with its sensitivity and adaptive capacity. It increases as the magnitude of climate change or sensitivity increases and decreases as adaptive capacity increases (OECD 2009).

Graphically, the concept of vulnerability has been represented as ultimately a balance between the system's adaptive capacity and the potential impacts to the system.

Utilities can take a progressive approach to assessing vulnerabilities by doing a wide-scope/low-detail assessment, followed by prioritization and a focused-scope/high-detail analysis, as depicted in Figures 7-3 and 7-4. A useful resource for the first step of low-detail multi-sector assessment is the vulnerability matrix developed by CDWR (USEPA Region 9 CADWR 2011).

Figure 7-4 Progressive approach to vulnerability assessment

An important consideration for assessing vulnerability is the varying geography of the utility's service area, supply sources (water and energy), disposal sites, critical infrastructure, and critical habitat areas providing ecosystem services to the utility. A vulnerability analysis is typically an array of individual analyses by sector and by geography as necessary.

Once vulnerabilities have been identified, the next step is to prioritize vulnerabilities so that the utility will be able to allocate resources during the planning process and when defining adaptation actions. Prioritizing vulnerabilities can be a political challenge, as explained earlier in this chapter, and should only be conducted with a full understanding of the priorities, objectives, and values of the stakeholders involved. However, this step helps focus limited resources on areas of greater importance.

Prioritization considerations typically involve

1. known vulnerabilities from recent extreme events;
2. risk associated with the vulnerabilities, understood as the impact multiplied by the probability of that impact; and
3. presence of multiple potential stressors, climate-related or not, that may decrease the adaptive capacity of the system (population growth, deforestation, significant land-use changes, loss of wetlands, etc.).

For the prioritized vulnerability areas, the utility should perform a detailed vulnerability and risk assessment and estimate impacts for areas identified as potential risks. Expected outcomes from a vulnerability assessment will depend on the sectors of the water resources and infrastructure system being evaluated but should include specific metrics. Vulnerability metrics may include the following:

- Water supply: reduced safe yield from a reservoir
- Water demands: increase summer agricultural demand
- Flooding: increase in expected flooded area under a 100-y storm
- Habitat: loss of habitat area
- Infrastructure: Number of key facilities flooded or impacted by a 100-y event

I. Water Demand

☐ *Are there major industries that require cooling/process water in your planning region?*

- As average temperatures increase, cooling water needs may also increase.
- Identify major industrial water users in your region and assess their current and projected needs for cooling and process water.

☐ *Does water use vary by more than 50% seasonally in parts of your region?*

- Seasonal water use, which is primarily outdoor water use, is expected to increase as average temperatures increase and droughts become more frequent.
- Where water use records are available, look at total monthly water uses averaged over the last five years (if available). If maximum and minimum monthly water uses vary by more than 25%, then the answer to this question is "yes".
- Where no water use records exist, is crop irrigation responsible for a significant (say >50%) percentage of water demand in parts of your region?

☐ *Are crops grown in your region climate-sensitive? Would shifts in daily heat patterns, such as how long heat lingers before night-time cooling, be prohibitive for some crops?*

- Fruit and nut crops are climate-sensitive and may require additional water as the climate warms.

☐ *Do groundwater supplies in your region lack resiliency after drought events?*

- Droughts are expected to become more frequent and more severe in the future. Areas with a more hardened demand may be particularly vulnerable to droughts and may become more dependent on groundwater pumping.

☐ *Are water use curtailment measures effective in your region?*

- Droughts are expected to become more frequent and more severe in the future. Areas with a more hardened demand may be particularly vulnerable to droughts.

☐ *Are some instream flow requirements in your region either currently insufficient to support aquatic life, or occasionally unmet?*

- Changes in snowmelt patterns in the future may make it difficult to balance water demands. Vulnerabilities for ecosystems and municipal/agricultural water needs may be exacerbated by instream flow requirements that are:
    1. not quantified,
    2. not accurate for ecosystem needs under multiple environmental conditions including droughts, and
    3. not met by regional water managers.

II. Water Supply

☐ *Does a portion of the water supply in your region come from snowmelt?*

- Snowmelt is expected to decrease as the climate warms. Water systems supplied by snowmelt are therefore potentially vulnerable to climate change.
- Where watershed planning documents are available, refer to these in identifying parts of your region that rely on surface water for supplies; if your region contains surface water supplies originating in watersheds where snowpack accumulates, the answer to this question is "Yes."
- Where planning documents are not available, identify major rivers in your region with large users. Identify whether the river's headwaters are fed by snowpack.

☐ *Does part of your region rely on water diverted from the Delta, imported from the Colorado River, or imported from other climate-sensitive systems outside your region?*

- Some imported or transferred water supplies are sources from climate-sensitive watersheds, such as water imported from the Delta and the Colorado River.

☐ *Does part of your region rely on coastal aquifers? Has salt intrusion been a problem in the past?*

- Coastal aquifers are susceptible to salt intrusion as sea levels rise, and many have already observed salt intrusion due to over-extraction, such as the West Coast Basin in southern California.

☐ *Would your region have difficulty in storing carryover supply surpluses from year to year?*

- Droughts are expected to become more severe in the future. Systems that can store more water may be more resilient to droughts.

☐ *Has your region faced a drought in the past during which it failed to meet local water demands?*

- Droughts are expected to become more severe in the future. Systems that have already come close to their supply thresholds may be especially vulnerable to droughts in the future.

☐ *Does your region have invasive species management issues at your facilities, along conveyance structures, or in habitat areas?*

- As invasive species are expected to become more prevalent with climate change, existing invasive species issues may indicate an ecological vulnerability to climate change.

III. Water Quality

☐ *Are increased wildfires a threat in your region? If so, does your region include reservoirs with fire-susceptible vegetation nearby which could pose a water quality concern from increased erosion?*

- Some areas are expected to become more vulnerable to wildfires over time. To identify whether this is the case for parts of your region, the California Public Interest Energy Research (PIER) Program has posted wildfire susceptibility projections as a Google Earth application at: http://cal-adapt.org/fire/. These projections are only the results of a single study and are not intended for analysis, but can aid in qualitatively answering this question. Read the application's disclaimers carefully to be aware of its limitations.

☐ *Does part of your region rely on surface water bodies with current or recurrent water quality issues related to eutrophication, such as low dissolved oxygen or algal blooms? Are there other water quality constituents potentially exacerbated by climate change?*

- Warming temperatures will result in lower dissolved oxygen levels in water bodies, which are exacerbated by algal blooms and in turn enhance eutrophication. Changes in streamflows may alter pollutant concentrations in water bodies.

☐ *Are seasonal low flows decreasing for some waterbodies in your region? If so, are the reduced low flows limiting the waterbodies' assimilative capacity?*

- In the future, low flow conditions are expected to be more extreme and last longer. This may result in higher pollutant concentrations where loadings increase or remain constant.

(Continued)

- ☐ *Are there beneficial uses designated for some water bodies in your region that cannot always be met due to water quality issues?*
  - In the future, low flows are expected decrease, and to last longer. This may result in higher pollutant concentrations where loadings increase or remain constant.
- ☐ *Does part of your region currently observe water quality shifts during rain events that impact treatment facility operation?*
  - While it is unclear how average precipitation will change with temperature, it is generally agreed that storm severity will probably increase. More intense, severe storms may lead to increased erosion, which will increase turbidity in surface waters. Areas that already observe water quality responses to rainstorm intensity may be especially vulnerable.

IV. Sea Level Rise

- ☐ *Has coastal erosion already been observed in your region?*
  - Coastal erosion is expected to occur over the next century as sea levels rise.
- ☐ *Are there coastal structures, such as levees or breakwaters, in your region?*
  - Coastal structures designed for a specific mean sea level may be impacted by sea level rise.
- ☐ *Is there significant coastal infrastructure, such as residences, recreation, water and wastewater treatment, tourism, and transportation) at less than six feet above mean sea level in your region?*
  - Coastal flooding will become more common, and will impact a greater extent of property, as sea levels rise. Critical infrastructure in the coastal floodplain may be at risk.
  - Digital elevation maps should be compared with locations of coastal infrastructure.
- ☐ *Are there climate-sensitive low-lying coastal habitats in your region?*
  - Low-lying coastal habitats that are particularly vulnerable to climate change include estuaries and coastal wetlands that rely on a delicate balance of freshwater and salt water.
- ☐ *Are there areas in your region that currently flood during extreme high tides or storm surges?*
  - Areas that are already experiencing flooding during storm surges and very high tides, are more likely to experience increased flooding as sea levels rise.
- ☐ *Is there land subsidence in the coastal areas of your region?*
  - Land subsidence may compound the impacts of sea level rise.
- ☐ *Do tidal gauges along the coastal parts of your region show an increase over the past several decades?*
  - Local sea level rise may be higher or lower than state, national, or continental projections.
  - Planners can find information on local tidal gauges at http://tidesandcurrents.noaa.gov/sltrends/sltrends_states.shtml?region=ca.

(Continued)

V. Flooding

☐ *Does critical infrastructure in your region lie within the 200-year floodplain? DWR's best available floodplain maps are available at: http://www.water.ca.gov/floodmgmt/lrafmo/fmb/fes/best_available_maps/.*

- While it is unclear how average precipitation will change with temperature, it is generally agreed that storm severity will probably increase. More intense, severe storms may lead to higher peak flows and more severe floods.
- Refer to FEMA floodplain maps and any recent FEMA, US Army Corps of Engineers, or DWR studies that might help identify specific local vulnerabilities for your region. Other follow-up questions that might help answer this question:
    1. What public safety issues could be affected by increased flooding events or intensity? For example, evacuation routes, emergency personnel access, hospitals, water treatment and wastewater treatment plants, power generation plants and fire stations should be considered.
    2. Could key regional or economic functions be impacted from more frequent and/or intense flooding?

☐ *Does part of your region lie within the Sacramento-San Joaquin Drainage District?*

- The SSJDD contains lands that are susceptible to overflows from the Sacramento and San Joaquin Rivers, and are a key focus of the Central Valley Flood Protection Plan. (http://www.water.ca.gov/cvfmp/program.cfm).

☐ *Does aging critical flood protection infrastructure exist in your region?*

- Levees and other flood protection facilities across the state of California are aging and in need of repair. Due to their overall lowered resiliency, these facilities may be particularly vulnerable to climate change impacts.
- DWR is evaluating more than 300 miles of levees in the San Joaquin and Sacramento Rivers Valleys and the Delta (http://www.water.ca.gov/levees/).

☐ *Have flood control facilities (such as impoundment structures) been insufficient in the past?*

- Reservoirs and other facilities with impoundment capacity may be insufficient for severe storms in the future. Facilities that have been insufficient in the past may be particularly vulnerable.

☐ *Are wildfires a concern in parts of your region?*

- Wildfires alter the landscape and soil conditions, increasing the risk of flooding within the burn and downstream areas. Some areas are expected to become more vulnerable to wildfires over time. To identify whether this is the case for parts of your region, the California Public Interest Energy Research Program (PIER) has posted wildfire susceptibility projections as a Google Earth application at: http://cal-adapt.org/fire/. These projections are the results of only a single study and are not intended for analysis, but can aid in qualitatively answering this question. Read the application's disclaimers carefully to be aware of its limitations.

VI. Ecosystem and Habitat Vulnerability

☐ *Does your region include inland or coastal aquatic habitats vulnerable to erosion and sedimentation issues?*

- Erosion is expected to increase with climate change, and sedimentation is expected to shift. Habitats sensitive to these events may be particularly vulnerable to climate change.

☐ *Does your region include estuarine habitats which rely on seasonal freshwater flow patterns?*

- Seasonal high and low flows, especially those originating from snowmelt, are already shifting in many locations.

(Continued)

- ☐ *Do climate-sensitive fauna or flora populations live in your region?*
  - Some specific species are more sensitive to climate variations than others.
- ☐ *Do endangered or threatened species exist in your region? Are changes in species distribution already being observed in parts of your region?*
  - Species that are already threatened or endangered may have a lowered capacity to adapt to climate change.
- ☐ *Does the region rely on aquatic or water-dependent habitats for recreation or other economic activities?*
  - Economic values associated with natural habitat can influence prioritization.
- ☐ *Are there rivers in your region with quantified environmental flow requirements or known water quality/quantity stressors to aquatic life?*
  - Constrained water quality and quantity requirements may be difficult to meet in the future.
- ☐ *Do estuaries, coastal dunes, wetlands, marshes, or exposed beaches exist in your region? If so, are coastal storms possible/frequent in your region?*
  - Storm surges are expected to result in greater damage in the future due to sea level rise. This makes fragile coastal ecosystems vulnerable.
- ☐ *Does your region include one or more of the habitats described in the Endangered Species Coalition's Top 10 habitats vulnerable to climate change (http://www.itsgettinghotoutthere.org/)?*
  - These ecosystems are particularly vulnerable to climate change.
- ☐ *Are there areas of fragmented estuarine, aquatic, or wetland wildlife habitat within your region? Are there movement corridors for species to naturally migrate? Are there infrastructure projects planned that might preclude species movement?*
  - These ecosystems are particularly vulnerable to climate change.

VII. Hydropower

- ☐ *Is hydropower a source of electricity in your region?*
  - As seasonal river flows shift, hydropower is expected to become less reliable in the future.
- ☐ *Are energy needs in your region expected to increase in the future? If so, are there future plans for hydropower generation facilities or conditions for hydropower generation in your region?*
  - Energy needs are expected to increase in many locations as the climate warms. This increase in electricity demand may compound decreases in hydropower production, increasing its priority for a region.

(Continued)

Source: Climate Change Handbook for Regional Water Planning, USEPA Region 9, CADWR, 2011

Figure 7-5 Matrix describing system vulnerabilities

In these examples, the metrics are specific and quantitative. Documents from WUCA describe several available tools that can be used in vulnerability assessments (WUCA, n.d.). Additionally, USEPA's creating resilient utilities resources (USEPA, n.d.), specifically the CREAT referenced in Chapter 5, are great starting points for the selection of appropriate tools for these type of analyses. The case studies in Chapter 10 and in the "Adaptation Case Study and Information Exchange" illustrate how utilities across the nation conduct different sector analyses.

Developing and Selecting Adaptation Strategies and Actions

In considering adaptation strategies, a utility should consult master planning, regional planning, and strategic planning documents to learn about relevant projects. Water, wastewater, and stormwater master plans, as well as capital improvement programs, often provide specific projects that can be aligned with a utility's climate change resilience goals. Asset management programs also have specific programmatic goals that can be aligned with a CAP as a source of adaptation measures. No matter the source—master plans, integrated resources plans, capital improvement plans (CIPs), or asset management—existing projects may need to be modified to better address adaptation goals. In some cases, however, entirely new projects may be required to achieve objectives established in a CAP.

For vulnerabilities that have been prioritized and evaluated quantitatively, specific impacts can be mitigated with CAP actions. Two challenges of developing a CAP are particularly exacerbated by the nature of climate change impacts:

1. Attempted solutions are as multi-sectorial as the impacts themselves. In comparison, a water supply plan will try to find sources of supply for a forecasted supply gap—all actions will be ultimately about managing demand or providing new supply. In a CAP, however, impacts may occur in supply sources, demands, ecosystems, flooding, hydropower generation, or other sectors, so the actions in a CAP often span several sectors of the water resources system, limited by the specific mission and core operating functions of the utility.
2. The uncertainty associated with climate change forecast make the quantification of "gaps" more challenging than in traditional planning. Risk and uncertainty are inherent in climate change planning and must be formally addressed in the development and evaluation of adaptation measures. Risk and uncertainty are discussed in Chapter 4 of this manual, and as well as in the document "Decision support planning methods: Incorporating climate change uncertainties into water planning" (WUCA 2010).

Even though the climate change impacts identified in the vulnerability assessment will include a degree of uncertainty, a "probabilistic gap" can be a good starting point to define the needs that will trigger adaptation actions. Actions may include projects or programs related to

- new facilities and infrastructure,
- green infrastructure,
- redesign,
- retrofit,
- operational changes,
- integrated operation,
- maintenance,

- water conservation and water use efficiency,
- habitat restoration, protection and expansion, and
- land use planning and management.

Figure 7-6 illustrates the concept of using vulnerabilities to define projects that address a specific need. Even if the magnitude of a need is uncertain, this process should identify actions that will potentially address that need.

Multi-sector climate change impacts are typically addressed by developing portfolios of actions. Existing literature demonstrates the value of multi-project (portfolio) approaches to addressing potential climate change impacts. To explore portfolio development and evaluation, refer to the Colorado River Basin study (USBOR 2012). Alternatively, some adaptation plans and policy documents need only define strategies to adapt to climate change and decrease the vulnerability of different sectors. This approach usually results in specific recommendations for the next planning steps to define specific projects and programs to address specific vulnerabilities. The Safeguarding California and Climate Change Adaptation Policy is an example of a comprehensive adaptation document that proposes strategies for later planning phases to develop projects (California Natural Resources Agency 2018).

Figure 7-7 shows the difference between a plan that provides strategies for each prioritized vulnerability area versus a plan that provides specific recommended actions. Individual projects can also be impacted by climate change, and the impact of climate change on the performance of those projects must be analyzed. This step is particularly important when projects have been characterized in other planning documents (master planning, feasibility studies, etc.) without evaluation of their potential performance under climate change conditions.

Projects that require evaluation can usually be evaluated with the same tools used to quantify impacts. Models can be adapted to incorporate proposed projects and programs. The benefits expected from different projects and programs can then be quantified, accounting for climate change impacts.

The chain of models concept presented in the WUCA document "Actionable Science in Practice" (WUCA 2015) shows the sequence of data, information, and models that are likely needed in a comprehensive evaluation of projects. Abundant literature discusses the use of general circulation model (GCM) outputs for quantitative analysis. Additionally, many case studies illustrate how to use that information.

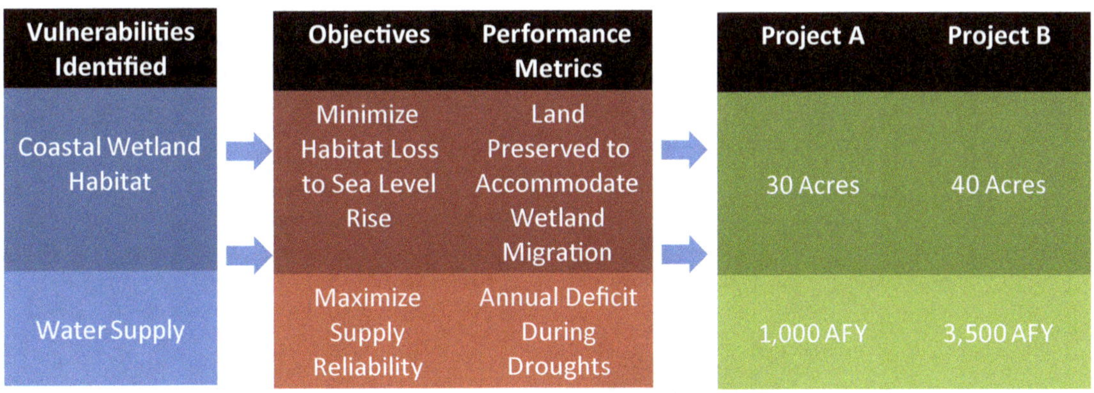

Source: Climate Change Handbook for Regional Water Planning, USEPA Region 9, CADWR, 2011

Figure 7-6 Vulnerabilities dictate objective of the CAP and the definition of actions to mitigate those vulnerabilities

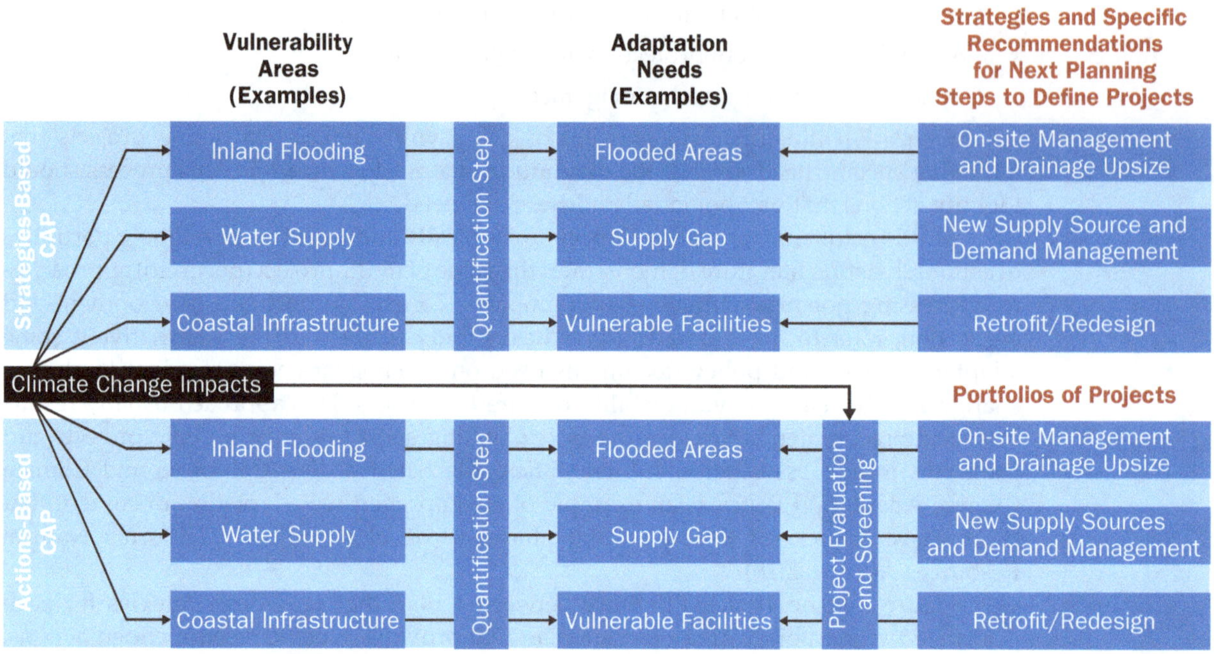

Figure 7-7 A plan can identify strategies and a path for future work to establish actions or establish actions after an analysis of how projects achieve the objectives of vulnerability mitigation

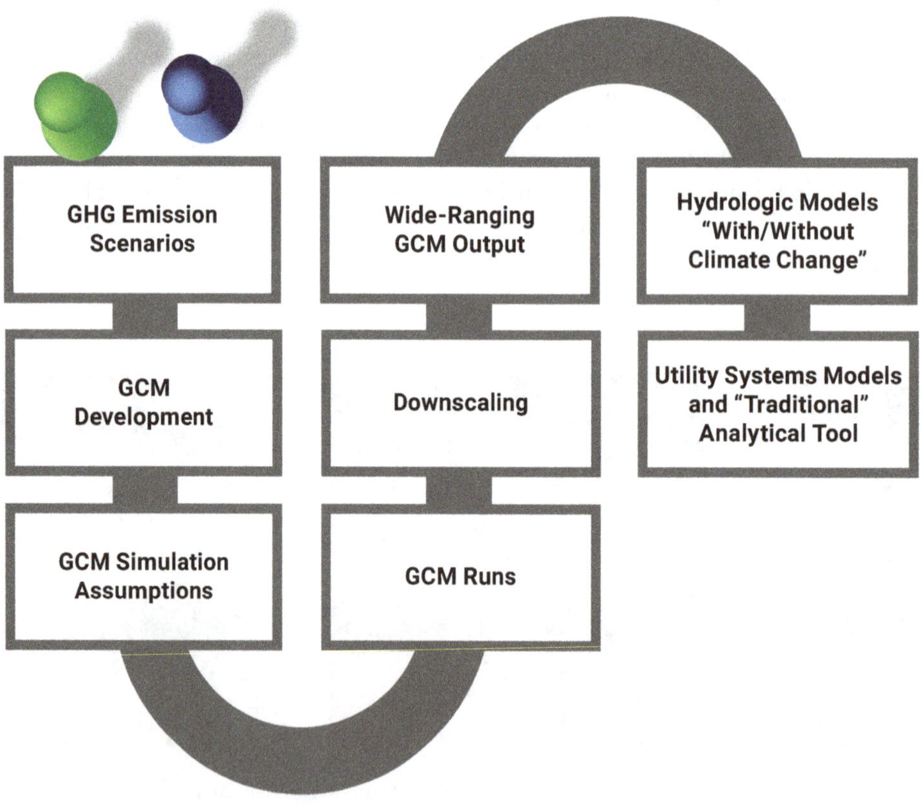

GCM—general circulation model, GHG—greenhouse gas

Figure 7-8 Sequence of steps to generate a climate change analysis

A utility should consider that projects can be evaluated with differing levels of sophistication, depending on the availability of information, tools, and resources, as shown in Figure 7-7. The CREAT is quantitative and includes relevant GCM predictions for different climatic variables. The tool is used by many agencies across the United States, a broad user base that provides an additional benefit for evaluation of adaptation measures.

Figure 7-8 provides a water resources example of the traditional sequence of modeling and analytical steps for evaluating projects and vulnerability. Most utilities will seek to obtain inputs for their analysis without needing to address the first five steps. This is possible in most cases. Utilities should consult with a climate change expert to identify the downscaled GCM outputs that are most appropriate for the type of analysis required.

Analysis requires many steps. Utilities will be interested in obtaining inputs for hydrologic models ("Wide-Ranging GCM Output" step) and should consult with a climate change expert to understand the origin of the selected inputs.

As discussed in Chapter 3, while downscaling outputs from GCMs is an initial step in most case studies, there are alternate approaches that recommend first defining the system's sensitivity to varying climatic variables. Garcia et al. (2014) describe a method in which downscaled data is only used to identify predicted futures under which a system has been determined to be vulnerable.

In the evaluation and selection of actions to include in an adaptation plan, risk and uncertainty are critically important to address formally. Chapter 4 discusses the management of risk and uncertainty in the planning process, as does the WUCA document "Actionable Science in Practice" (WUCA 2015). Some case studies in Chapter 10 include examples of methods to address climate change uncertainty in decision-making. Scenario planning (Groves et al. 2013), decision-scaling (Garcia et al. 2014), adaptive management, and robust decision making (RAND 1969) are methods that have been used to define strategies for adaptation.

In selecting actions for implementation, the costs of evaluated actions should be assessed and included as part of the selection criteria for the plan's recommended portfolio of actions. While costs should not define the final choices—particularly considering the multiple negative externalities of climate change—political and community support will be easier to achieve after a formal assessment of costs. In many cases, cost estimates may already be available from master planning efforts, a CIP, or other planning documents. When no costs are available for specific projects or programs, conceptual cost estimates should be adequate for a CAP. An example of a conceptual level cost estimate for recommended options can be found in the Equinox Center *San Diego Water Sources: Assessing the Options* (2010).

CAP IMPLEMENTATION

The final component of a CAP is setting up an implementation strategy to plan for actions that will mitigate the prioritized vulnerabilities.

As illustrated in Figure 7-7, some CAPs may only include strategies to deal with the specific vulnerabilities in each sector. For those strategy-based plans, the implementation path should identify the specific planning mechanisms to evaluate projects and programs that can directly address vulnerabilities. For plans that include a recommended portfolio of actions, the implementation strategy can identify a timeline for actions. In both strategy-based plans and action-based plans, the implementation strategy should include a specific monitoring plan. The variables to monitor should match the variables that are critically linked to the vulnerabilities in each sector (annual snowpack, stream flows, number of storm events over a certain depth, annual rainfall, sea level, etc.). The monitoring plan should add specificity about the running averages (transient) that should be computed as indicators to eliminate the natural annual variability, as discussed in the mitigation section in this chapter. USEPA publishes specific climate change indicators at national and regional levels: Climate Change Indicators in the United States (USEPA 2016).

Monitoring specific variables is particularly important to define the timing for implementation of actions under an adaptive management framework. Adaptive management has been extensively applied to water resources. The Colorado River Basin study lays out an adaptive management process for implementation of projects under a preferred portfolio. Williams and Brown (2012) explain in detail the concepts of adaptive management.

An additional element of CAP implementation should be continued engagement with stakeholders, partner agencies, and the scientific community, as discussed in Chapter 8.

REFERENCES

Alliance for Water Efficiency. Water Conservation Tracking Tool. https://www.allianceforwaterefficiency.org/resources/topic/water-conservation-tracking-tool. (accessed April 9, 2020).

California Natural Resources Agency, 2009. 2009 *California Climate Adaptation Strategy*. https://resources.ca.gov/CNRALegacyFiles/docs/climate/Statewide_Adaptation_Strategy.pdf (accessed April 9, 2020).

Christine, Z.F., Kulkarni, K. Polgar, S. Schneider, M. Webster, S.S. Greenhouse Gas Mitigation: A Guide for Small Municipal Utilities. Santa Barbara, CA: Donald Bren School of Environmental Science and Management. 2005. https://www.bren.ucsb.edu/research/documents/co2_final.pdf (accessed April 8, 2020).

Climate Change Handbook for Regional Water Planning, EPA Region 9, CADWR, 2011.

Equinox Center. 2010. *San Diego's Water Sources: Assessing the Options*. https://energycenter.org/sites/default/files/San_Diego_Water_Sources_Assessing_the_Options_July_2010_FBI.pdf (accessed April 9, 2020).

García, L.E., Matthews, J.H. Rodriguez, D.J., Wijnen, M., DiFrancesco, K.N. Ray, P. 2014. Be-yond Downscaling: A Bottom-Up Approach to Climate Adaptation for Water Resources Management. AGWA Report 01. Washington, DC: World Bank Group.

Gleick P., Stein, R. and Edelson, N. editors. 2011. Scanning the conservation horizon: a guide to climate change vulnerability assessment. National Wildlife Federation, Washington, D.C.

Greenhouse Gas Protocol. http://www.ghgprotocol.org/. (accessed April 9, 2020).

Groves et al, 2013. Groves, David G.; Fischbach, Jordan R.; Bloom, Evan; Knopman, Debra; Keefe, Ryan. Adapting to a Changing Colorado River: Making Future Water Deliveries More Reliable Through Robust Management Strategies. Santa Monica, CA: RAND Corporation, 2013. https://www.rand.org/pubs/research_reports/RR242.html.

Heather, C., Allen, L. Heberger, M. and Wilkinson, R. Implications of Future Water Supply Sources for Energy Demands. Pacific Institute (2012). http://pacinst.org/publication/wesim/. (accessed April 9, 2020).

Huxley, D.E., Bellamy, W.D. Sathyanarayan, P. Ridens, M. 2009. "Greenhouse Gas Emission Inventory and Management Strategy for Water Utilities." The Water Research Foundation (formerly AwwaRF) (WRF) and California Urban Water Agencies (CUWA). 2009. Englewood, Colo: CH2M HILL.

IPCC, 2007: *Climate Change 2007: Impacts, Adaptation and Vulnerability. Contribution of Working Group II to the Fourth Assessment Report of the Intergovernmental Panel on Climate Change*, M.L. Parry, O.F. Canziani, J.P. Palutikof, P.J. van der Linden and C.E. Hanson, Eds., Cambridge University Press, Cambridge, UK, 976pp. https://www.ipcc.ch/report/ar4/wg2/ (accessed April 9, 2020).

IPCC, 2013: *Climate Change 2013: The Physical Science Basis. Contribution of Working Group I to the Fifth Assessment Report of the Intergovernmental Panel on Climate Change* [Stocker, T.F., D. Qin, G.-K. Plattner, M. Tignor, S.K. Allen, J. Boschung, A. Nauels, Y. Xia, V. Bex and P.M. Midgley (eds.)]. Cambridge University Press, Cambridge, United Kingdom and New York, NY, USA, 1535 pp. https://www.ipcc.ch/report/ar5/wg1/ (accessed April 9, 2020).

Norman, C.D. The Delphi Method: An Experimental Study of Group Opinion. RAND Corp, Santa Monica, California. June 1969. https://apps.dtic.mil/docs/citations/AD0690498. (accessed April 9, 2020).

OECD – Organisation for Economic Co-operation and Development 2009: *Integrating Climate Change Adaption into Development Co-operation*. https://www.oecd.org/env/cc/44887764.pdf (accessed April 9, 2020).

POLIS Water Sustainability Project. WaterSmart Tool Kit. http://www.poliswaterproject.org/toolkit#calculator. (accessed April 8, 2020).

Raucher, R. et al. 2014. Effective Climate Change Communication for Water Utilities. Water Research Foundation, Denver.

TCR (The Climate Registry). http://www.theclimateregistry.org/tools-resources/reporting-protocols/general-reporting-protocol/ (accessed April 9, 2020).

The California Governor's Office for Emergency Services. March 2020. *California Adaption Planning Guide*.https://www.caloes.ca.gov/HazardMitigationSite/Documents/APG2-FINAL-PR-DRAFT Accessible.pdf (accessed April 9, 2020).

US Global Research Program. 2014. National Climate Assessment. https://nca2014.globalchange.gov/report#submenu-report-response-strategies (accessed April 9, 2020).

USBR (US Bureau of Reclamation). 2012. Colorado River Basin Water Supply and Demand Study. https://www.usbr.gov/lc/region/programs/crbstudy.html (accessed April 9, 2020).

USEPA (US Environmental Protection Agency). 2016. *Climate Change Indicators in the United States*. Fourth edition. https://www.epa.gov/sites/production/files/2016-08/documents/climate_indicators_2016.pdf (accessed April 9, 2020).

USEPA. 2013. Energy Efficiency in Water and Wastewater Facilities: A Guide to Developing and Implementing Greenhouse Gas Reduction Programs. https://www.epa.gov/sites/production/files/2015-08/documents/wastewater-guide.pdf (accessed April 8, 2020).

USEPA. EPA Center for Corporate Climate Leadership. https://www.epa.gov/climateleadership (accessed April 9, 2020).

USEPA. 2014. The EPA Corporate GHG Goal Evaluation Model: A Model for Benchmarking GHG Reductions and Evaluating Corporate Climate Performance. https://www.epa.gov/sites/production/files/2015-07/documents/epa_ghg_goal_model_users_guide_8.14_final.pdf (accessed April 9, 2020).

USEPA Energy Resources for State, Local and Tribal Governments http://www3.epa.gov/statelocalclimate/documents/pdf/wastewater-guide.pdf. (accessed April 9, 2020).

USEPA Creating Resilient Water Utilities (CRWU). https://www.epa.gov/crwu (accessed April 9, 2020).

USEPA Creating Resilient Water Utilities (CRWU) Case study and Information Exchange. https://epa.maps.arcgis.com/apps/MapSeries/index.html?appid=03d35ca84b5944f8b3ab59bf3a981462 (accessed April 9, 2020).

USEPA CREAT Risk Assessment Application for Water Utilities. https://www.epa.gov/crwu/creat-risk-assessment-application-water-utilities (accessed April 9, 2020).

Water Utility Climate Alliance. https://www.wucaonline.org/ (accessed April 9, 2020).

WESTWeb. Water-Energy Sustainability Tool http://west.berkeley.edu/. (accessed April 9, 2020.)

Williams, B. K., and Brown, E. D. 2012. Adaptive Management: The U.S. Department of the Interior Applications Guide. Adaptive Management Working Group, U.S. Department of the Interior, Washington, DC.

WRF (Water Research Foundation). 2009. Greenhouse Gas Emission Inventory and Management Strategy Guidelines for Water Utilities.

Chapter 8

Stakeholder Involvement and Public Participation

For purposes of this chapter, stakeholder involvement and public participation are both addressed as "public participation." This chapter (1) defines participation and its applicability to climate action planning (CAP) (2) provides a framework for developing a participation plan at all phases of the CAP process, and (3) provides the tools necessary to monitor and evaluate the effectiveness of the participation process.

WHAT IS PARTICIPATION?

Public participation is about more than just providing information to the public. When decision makers and stakeholders interact, they make better decisions (Creighton 2005). Ultimately, public participation includes a range of strategies that include different levels of engagement based on the need for input, understanding, and action. This range of strategies has been mapped as a spectrum, as shown in Table 8-1, by the International Association of Public Participation (IAP2). Each level of participation is defined by specific goals and carries specific promises to the public about how their participation will affect the process.

This spectrum is based on a set of core values defined by the IAP2:

- The public should have a say in decisions about actions that affect their lives.
- Public participation includes the promise that the public's contribution will influence the decision.
- The public participation process communicates the interests and meets the process needs of participants.

- The public participation process actively seeks out and facilitates the involvement of those potentially affected.
- The public participation process involves participants in defining how they participate.
- The public participation process provides participants with the information they need to participate in a meaningful way.
- The public participation process communicates to participants how their input affected the decision.

WHY IS PUBLIC PARTICIPATION IMPORTANT TO CLIMATE ACTION PLANNING?

Public participation is important to climate action planning for a variety of reasons. Many climate action planning issues directly impact the public, such as:

- inability of existing water supply sources to meet future demand,
- development of scenarios to test how best to meet future water supply demand,
- investment in new water supply sources,
- emergency planning for changes in water supplies and natural disasters,
- identification of water-use behaviors that may need to change, and
- rate changes to incentivize behavior change or to fund new infrastructure.

Table 8–1 IAP2 spectrum of public participation

	Inform	Consult	Involve	Collaborate	Empower
Public Participation Goal	To provide the public with balanced and objective information to assist them in understanding the problems, alternatives, and/or solutions.	To obtain public feedback on analysis, alternatives, and/or decisions.	To work directly with the public throughout the process to ensure that public issues and concerns are consistently understood and considered.	To partner with the public in each aspect of the decision, including the development of alternatives and the identification of the preferred solution.	To place final decision-making in the hands of the public.
Promise to the Public	We will keep you informed.	We will keep you informed, listen to and acknowledge concerns, and provide feedback on how public input influenced the decision.	We will work with you to ensure that your concerns and issues are directly reflected in the alternative developed and provide feedback on how public input influenced the decision.	We will look to you for direct advice and innovation in formulating solutions and incorporate your advice and recommendations into the decision to the maximum extent possible.	We will implement what you decide.
Example Tools	Fact sheets, websites,ebsites open houses	Public comment, focus groups, surveys, public meetings	Workshops, deliberate polling	Citizen advisory committees, consensus-building, participatory decision-making	Citizen juries, ballots, delegated decisions

Source: IAP2 Spectrum of Public Participation 2016

Public participation shares information about complex issues, alternatives, and trade-offs. Furthermore, public input can lead to more informed decisions. Within climate action planning, both adaptation and mitigation strategies affect stakeholders, customers, and members of the public. Climate adaptation often involves changes to water systems and water use behavior. Similarly, mitigation measures—implementing new technologies and renewable energies, retrofitting older equipment to improve efficiency, and changing management practices or consumer behavior—also impact the public.

OPPORTUNITIES FOR PUBLIC PARTICIPATION IN CLIMATE ACTION PLANNING

To address these issues, CAPs generally include five planning milestones, each with specific opportunities for public participation. These five milestones are (1) initiate, (2) research, (3) plan, (4) implement, and (5) monitor/review (ICLEI 2015). These milestones can occur out of sequence, as climate action planning is an iterative process.

During the five milestones of climate action planning, the public participation spectrum will help teams evaluate the need for participation and define the level of involvement appropriate for that need. This type of consideration ensures that public participation

- builds opportunities for clearer understanding of issues;
- invites the type of input that can help to develop a well-informed strategy of how water will be supplied, managed, delivered, and used;
- allows the public to influence decisions that affect their lives;
- demonstrates government accountability in policy, regulatory, and environmental processes; and
- provides an opportunity to develop public understanding of final decisions.

HOW TO ENCOURAGE PUBLIC PARTICIPATION IN EACH MILESTONE

This section describes considerations specific to the various planning milestones with respect to public participation. There are four tasks that are universal to all milestones: (1) developing a communications and engagement plan specific to the current planning phase; (2) identifying stakeholders specific to that planning phase; (3) defining and inviting leadership participation; and (4) facilitating staff meetings and project management. This section focuses on the first two items that directly relate to public participation.

Developing a Communications and Engagement Plan Specific to Each Milestone

Ideally, planning efforts include a communications plan and an engagement plan that work together. Communications plans generally provide a situational analysis to identify key factors that are causing the need for communications. This is followed by audience analysis, opportunities and constraints analysis, core messaging and branding, communications strategies, resource identification, and evaluation methods.

Engagement plans identify where the decision points are within the planning effort and the types of input that would help inform those decisions and allow public participation. These plans benefit from using the public participation spectrum introduced earlier in this chapter. A range of involvement strategies are detailed in a reference toolbox at the end of this chapter. The public's need for information, as well as publicity opportunities, can be incorporated into the communications plan.

Stakeholder Identification Specific to Each Milestone

Throughout the planning process, there will be internal and external stakeholders ("the public") that are relevant to the decisions being considered and implemented. The relevant stakeholders differ between milestones of the planning process and will change due to factors such as election cycles; leadership changes in community-based organizations, institutions, and businesses; turnover within communities as residents leave and new residents establish themselves; and property sales. The following checklists provide guidance on the stakeholders to engage throughout the planning process.

Internal stakeholders. Internal stakeholders are those within the planning agencies. They may be responsible for managing climate change impacts on infrastructure, public communications, natural resources, economic development, and emergency services. Examples of potential internal stakeholders include

- the management team,
- staff,
- related departments, and
- the board of directors and elected officials.

External stakeholders. External stakeholders exist outside the organization or planning agency. Examples of potential external stakeholders include

- local governments, regional planning organizations, other levels of government;
- community planning groups and other community-based organizations;
- universities, colleges, and schools;
- local businesses, commercial and industrial centers;
- non-profit organizations, professional associations, industry associations;
- insurers;
- utilities;
- researchers;
- environmental justice groups;
- customers/rate payers; and
- property owners/developers.

Milestones for Preparing a CAP

Milestone 1: Initiate. The first step of preparing a CAP is to establish the overall vision for the process and the level of support needed. This usually requires evaluating the context for the CAP by:

- developing a communications plan and engagement plan specific to the milestone,
- identifying potential internal and external stakeholders,
- building a climate change action team, and
- identifying a climate change action champion.

Milestone 2: Research. During this phase, the climate change action team works to conduct a risk and vulnerability assessment and identify adaptation and mitigation methods. Research usually includes

1. coordinating with internal stakeholders via formal staff meetings and
2. conducting
   - special workshops and ad hoc committees,
   - informal discussions,

- interviews,
- tours,
- review of government records and media archives, and
- surveys and

3. engaging NGOs, non-profits, businesses, institutions, and universities, as relevant to the project.

Milestone 3: Plan. During this phase, the climate action team tests the adaptation and mitigation methods to refine the plan. The final plan typically includes short- and long-term goals and actions. Participation in this phase focuses on testing tradeoffs of impacts and mitigation and adaptation measures. Methods typically include ways for people to evaluate and prioritize options. Involvement methods include

- workshops,
- surveys and polls,
- presentations, and
- partnership identification and development.

Milestone 4: Implement. During this phase, the CAP moves out of planning mode and into realization mode. Strong consideration should be given to the role of marketing so the CAP campaign can report progress to stakeholders, orient new stakeholders to the existing plan, and build relationships and resources needed for implementation. The implementation program may include convening a public monitoring team as well as changing behaviors, such as water use behavior, as related to the CAP. For behavior changes, community-based social marketing can reinforce desired behaviors and reduce barriers to those behaviors. This works through four main phases: research, design, engagement, and evaluation. The outreach process for this type of engagement pulls from a variety of social strategies, as shown in Table 8-2.

Milestone 5: Monitor/review. During this last milestone, CAP teams track the plan's progress, collect new information, update the plan, and communicate the plan's status with the community. To ensure accountability, the CAP team must develop a monitoring process and conduct an internal review. This review process should track CAP progress

Table 8-2 Community-based social marketing strategies

Strategy	Description
Commitments	Commitments demonstrate intent to take an action. They are strong motivators to the person making the commitment and to others who associate with that person. Commitments often start small, then build, and are especially effective when made in public.
Prompts	Prompts remind people to make the desired behavior change. This reminder comes as close as possible to the moment when they take that action.
Social Diffusion	Social diffusion speeds up adoption of the desired behavior when local community leaders demonstrate the behavior to fellow community members.
Incentives	Incentives encourage behavior change by demonstrating the benefits of a behavior, such as social recognition and monetary rewards.
Communications	Communications use a range of vehicles—print, digital, environmental, and peer-to-peer—to deliver the message.
Social Norms	Social norms to set and reinforce an ethic particular to people who identify with a certain group.
Convenience	Convenience is how easy it is for a community member to act on the request to change their behavior. This approach works to remove barriers (physical/structural, financial, policy, social) to make the actions easier to do.

by considering implementation and the effectiveness of actions. The following questions should be considered:

- What planned actions have been successfully implemented?
- What achievements were gained?
- What barriers were overcome?
- What barriers still exist?
- What are some possible solutions to solve barriers to climate action planning?

After analyzing the CAP's progress, the CAP team should work to determine whether new information exists and to update the existing plan with new research content. Lastly, the CAP team should continue with its communication efforts to participating stakeholders, team members, and the public. The IAP2 Public Participation Toolbox, included below, provides communication methods to inform interested groups. It serves as a comprehensive guide with techniques to share information, collect and compile input, and bring people together. This toolbox is a key component of the Foundations for Public Participation course taught by IAP2 and very relevant to how the CAP team should communicate and operate. The toolbox provides a step-by-step guidance for effective public participation in the CAP development process.

In addition, the CAP team should consider the following concerns:

- Internal goals for the public participation process
- The public's goals for the participation process
- Compliance with regulatory laws/requirements
- Public acceptance of final decisions
- Support of local officials
- Public satisfaction with the decision process
- Adequacy of final decision in incorporating public concerns

REFERENCES

Creighton, J. L. (2005). The Public Participation Handbook: Making Better Decisions Through Citizen Involvement. San Francisco: Jossey-Bass.

IAP2-International Association of Public Participation (2016). IAP2: Good public participation results in better decisions. http://www.iap2.org/.

ICLEI (Local Governments for Sustainability). Changing Climate, Changing Communities: Municipal Climate Adaptation Guide and Workbook. Retrieved from http://www.icleicanada.org/images/icleicanada/pdfs/GuideWorkbookInfoAnnexes_WebsiteCombo.pdf.

Krishnaswamy, A. (2012). Strategies and Tools for Effective Public Participation in Natural Resource Management. Journal Of Ecosystems And Management, 13(2). http://jem.forrex.org/index.php/jem/article/view/124.

McKenzie-Mohr, D. and Smith, W. 1999 . *Fostering Sustainable Behavior: An Introduction to Community-Based Social Marketing.* Gabriola Island, B.C.: New Society Publishers.

OECD (2015), Stakeholder Engagement for Inclusive Water Governance, OECD Publishing, Paris. DOI: http://dx.doi.org/10.1787/9789264231122-en.

United Nations Environment Programme. (2016). Climate Change Mitigation. Retrieved November 30, 2016, from http://www.unep.org/climatechange/mitigation/.

Chapter 9

Estimating the Cost of a Climate Action Plan

This chapter contains discussion on the following topics:

- Components of a climate action plan (CAP) cost estimate
- Tools available for CAP estimates
- Cost estimates from published CAPs
- Sources of funding

ESTIMATING THE COST OF A CLIMATE ACTION PLAN

The key to estimating the cost of a CAP for water utilities is the vulnerability assessment. Vulnerability is defined as "the degree to which a system is susceptible to, or unable to cope with, adverse effects of climate change, including climate variability and extremes" (Snover et al. 2007). Vulnerability assessments are used to evaluate possible threats to a system and to determine actions that can mitigate those effects.

The more vulnerable the utility, the more expensive adaptation will be. Investment in adaptation is directly related to how climate change will affect that geographical region and damage that is likely to occur given the utility's current resilience.

Vulnerability assessments fall into two categories: top-down and bottom-up (USEPA 2011). In a top-down approach, the vulnerability assessment is guided by climate models. These models are based on a combination of climate projections, historic hydrology, and paleoclimate records. This approach produces outputs using a global climate model that generates climate projections for the local watershed. By contrast, a bottom-up approach is guided by a utility's knowledge of the water system, including climate trends in the utility's region and the vulnerability of system components.

Vulnerability assessments for a water utility CAP can be divided into the following phases:

1. Determine the climate change threats to which the utility is susceptible.
2. Estimate the probability that a threat will occur.
3. Determine the repercussions of those threats for the water utility (impact on water quality, infrastructure, etc.).
4. Evaluate mitigation and adaptation measures.

Determine the Threats to the Utility

Before a utility can begin preparing a CAP that addresses adaptation strategies, it must first understand how climate change will present in that utility's area. Chapter 3 describes how climate change is expected to manifest across regions in the United States, Canada, and Mexico. Figure 9-1 summarizes expected regional climate change impacts to drinking water utilities (AMWA, 2009).

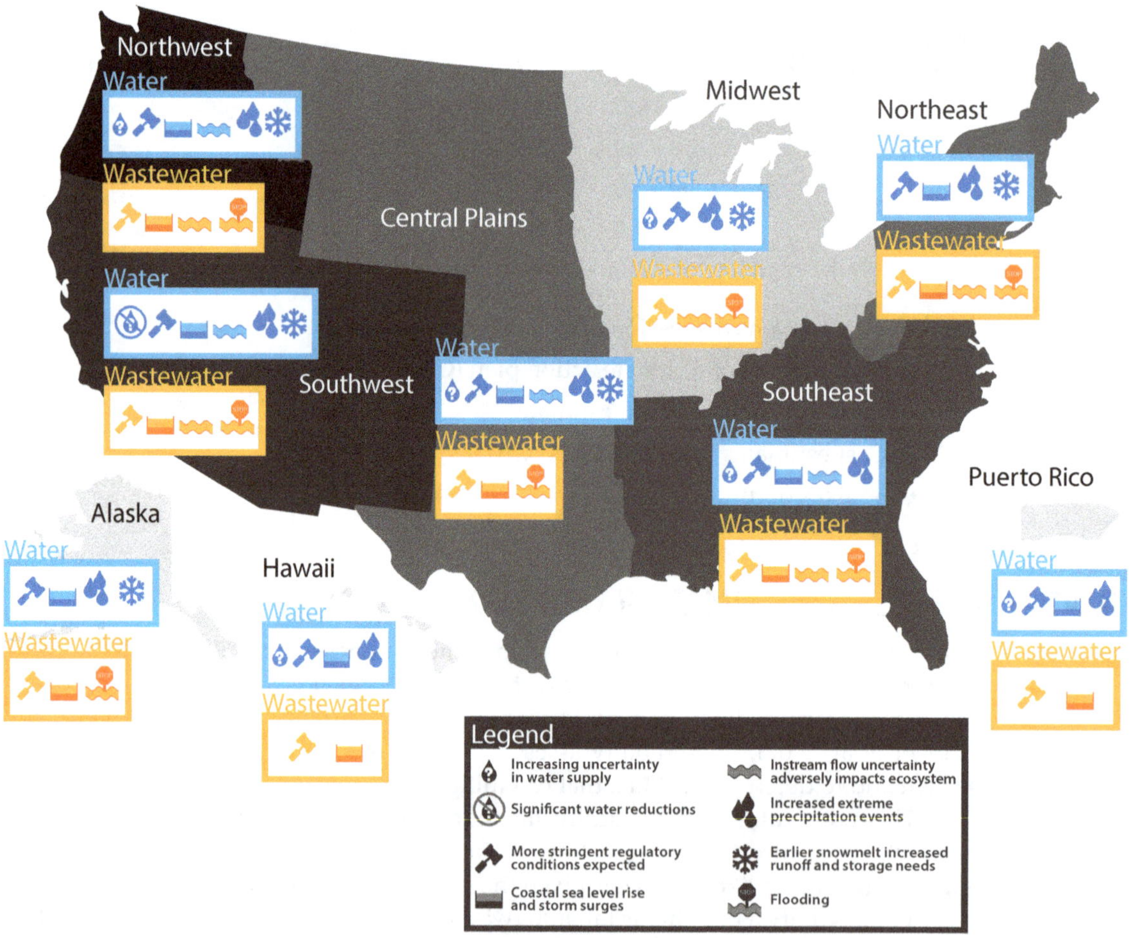

Source: NACWA and AMWA, Confronting Climate Change: An Early Analysis of the Cost of Climate Adaptation, 2009

Figure 9-1 Impacts of climate change on regions of the United States

In a top-down approach, this data can be provided by agencies that collect regional climate data and provide climate projection models for different parts of the country. One source is the US Environmental Protection Agency (USEPA), which provides climate projection data including annual total precipitation, annual average temperature, precipitation intensity for a 100-year storm, and sea level rise for 2035 and 2060. (EPA).

The National Oceanic and Atmospheric Administration (NOAA) provides a similar summary of past climate data. Information regarding climate change projections for regions across the United States can be obtained from the National Climate Assessment.

In a bottom-up approach, climate projection data is typically not used as the basis for the assessment. Rather, a utility will analyze data collected over a period of years to determine the number and severity of extreme weather events in the past. Common considerations include the following:

- Was the utility able to remain fully operational after the extreme weather event?
- If there was a shutdown, how much time elapsed before the utility was able to return online?
- Was the shutdown caused by a problem that occurred during previous extreme weather events?
- Did repairs to the system make it more resilient, or was the system repaired to its previous state?

Such questions help a utility determine its resilience to probable threats and pinpoint vulnerable assets.

Estimate the Probability That a Threat Will Occur

While it is not possible to attribute an individual weather event to climate change, climate projection models can estimate the risk of a specific extreme weather event in an area. The National Climate Assessment has combined data from several climate models and scenarios to provide detailed projections for the likelihood of certain weather events due to climate change (US Global Change).

USEPA provides a brief assessment of the probability of specific threats in various regions of the United States in "*Adaptation Strategies Guide for Water Utilities*" (2015). This guide indicates how strongly climate change effects such as drought, floods, and water quality deterioration will affect utilities in various regions across the United States, with additional information available for coastal regions. The guide is especially useful for utilities attempting to gauge their risk of experiencing extreme weather events caused by climate change.

Using these guides is part of a top-down approach, as utilities using bottom-up approaches may instead base their evaluation on existing data. However, for the purposes of a CAP, utilities using a bottom-up approach should consider these future probabilities as well as estimates from existing data.

Determine the Severity and Repercussions of Threats to the Water Utility

The AWWA Research Foundation summarizes the climate change repercussions water utilities can face in "Climate Change and Drinking Water, 2008." These repercussions include the following.

- Problems with water resources—Climate models agree that climate change will affect the timing and intensity of precipitation, as well as the timing of snowpack melt. Winter precipitation is anticipated to increase in intensity, while warmer temperatures are expected to cause earlier snowmelt. This may lead to insufficient water supplies during hot summer months. Furthermore, some areas will experience sustained droughts.

- Water quality issues—Periods of drought followed by heavy rainfall can increase the nutrient load entering source waters. High levels of nitrates and phosphates create an ideal environment for algal growth and can result in algal toxins as well as odor and taste concerns. Coastal areas that rely on groundwater may face saltwater intrusion due to rising sea levels.
- Infrastructure—Areas with increased rainfall intensity and frequency may experience increased turbidity and sedimentation, which may put a strain on facilities treating surface water. Flooding may also damage infrastructure both above and below ground.

Identifying how much an extreme weather event will affect an individual utility is even less certain than estimating the event's likelihood. The degree of damage can depend on both the severity of the event and the resilience of the utility's infrastructure. For example, a water utility in the Southeastern United States is less susceptible to high flow events and flooding than a utility in the Midwest, according to Adaptation Strategies Guide for Water Utilities (USEPA). However, a Midwestern utility that anchors fuel and chemical tanks and constructs levees may be considered more resilient to intense rain and flooding events than a comparable utility in the Southeast that has not implemented such resilience measures.

In "Adaptation Strategies Guide for Water Utilities," USEPA proposes that a utility document the threshold conditions for critical assets that can be negatively impacted by climate change. The conditions should be included in the vulnerability assessment regardless of whether a top-down or bottom-up approach is used. According to USEPA, "these thresholds can be determined through review of event and performance history, modeling of system performance, or inspection of assets. For example, the elevation of coastal facilities, combined with precipitation totals could define the flood stage or combination of sea level rise and storm intensity as thresholds for flooding."

Evaluate Mitigation and Adaptation Measures and Rate Their Effectiveness

"Adaptation Strategies Guide for Water Utilities" provides a detailed series of adaptation options water utilities can undertake to address the impacts of climate change. These measures are associated with a relative cost and are grouped into the following categories, with examples.

- Ecosystem and Land Use—Preserve vegetated land cover, manage stormwater runoff, implement green infrastructure, use natural ecosystems to reduce storm surge impacts or buffer against floods.
- Modeling—Model future water quality changes and groundwater conditions, analyze future sea level rise and storm surges, project runoff and future water supply.
- Monitoring—Monitor the integrity and resilience of current infrastructure, monitor weather conditions and the frequency and intensity of extreme events such as flooding, monitor surface water conditions and reservoir water quality.
- New Construction—Construct flood barriers to protect existing infrastructure, relocate utility infrastructure to higher elevations that are out of the flood plain, increase water storage capacity.
- Planning—Develop and maintain emergency response plans, insure utilities for a financial buffer in the event of an emergency, develop coastal restoration plans (if applicable), identify and secure vulnerable facilities.

- Repair and Retrofit—Increase treatment capabilities in the event that water quality is compromised, increase installation of backflow prevention devices, implement saltwater intrusion barriers, implement procedures for post-emergency repairs.
- System and Energy Efficiency—Introduce energy efficiency measures, introduce water recycling, practice conjunctive use of ground and surface water (using surface water and storing ground water during wet years and accessing stored ground water during dry years when surface water supply is low).
- Water Demand and Use—Encourage customers to implement water conservation measures, encourage water conservation in power plants, model and reduce water demand for agricultural and irrigation purposes.

Utilities that have already determined the threats they face may examine these options and determine the best adaptation strategies within their budgets to protect against these threats.

TOOLS TO ESTIMATE THE COST OF A CLIMATE ACTION PLAN

Establishing a vulnerability assessment may seem daunting, especially given the many sources of information. Fortunately, USEPA's CREAT allows water utilities to evaluate potential impacts of climate change on their utilities and assess potential adaptive measures and their costs. Created as a part of USEPA's Creating Resilient Water Utilities initiative, CREAT contains pre-loaded data relevant to wastewater and drinking water utility assets across the United States as well as customizable adaptation strategies. By consolidating data from multiple sources into a single tool, CREAT guides utilities to the data sets that are most appropriate for use in a vulnerability assessment.

CREAT incorporates both regional and local data to accurately assess the climate change threats and potential economic and environmental impacts that utilities face. Once the utility has decided on the level of action it wishes to undertake to protect against climate change impacts, CREAT will generate a report that provides climate data, risk reduction methods, and a cost estimate for implementing specific initiatives. This tool provides the flexibility to evaluate the costs and effectiveness of various adaptation options. Utilities can download and begin using CREAT at no cost by visiting USEPA's Creating Resilient Water Utilities webpage.

Table 9-1 summarizes the costs that example utilities have incurred in the implementation of a CAP, as well as their adaptation strategies. In most cases, climate change adaptation measures are part of larger CAPs that include additional measures such as establishing GHG inventories and reducing a city's carbon footprint. Budget estimates are reflected as such.

SOURCES OF FUNDING

Grants, and their eligibility requirements, are constantly evolving. This list is meant as a static guide to federal grants that are available to fund CAPs for water utilities.

Drinking Water State Revolving Fund

The Drinking Water State Revolving Fund (DWSRF) is a financial assistance program implemented in 1996 as an amendment to the Safe Drinking Water Act. USEPA allocates funding (provided as loans) to each state based on assessments of its drinking water infrastructure. To be eligible for a grant from the DWSRF, the utility must use the funding for infrastructure improvements related to source water, treatment, distribution, storage, new systems, or the consolidation of systems. Most infrastructure-related work in the execution of a CAP will meet the criteria for a DWSRF loan (USEPA).

Table 9-1 Cost estimates from published CAPs

City/Utility	Climate Change Impacts to Water Supply	Adaptive Measures in Place	Year CAP Created
New York City Department of Environmental Protection	Decreased quantity of water supply	Developing alternative auxiliary water sources such as groundwater	2008
Population served: 9 million + tourists Main Water Source: Surface Budget estimate: $1.5 million (for the development of a comprehensive strategy)		Backing up systems via interconnections with pipes in New Jersey, New York City, Connecticut, and Long Island	
		Using reservoir models and other tools to balance flows and releases to improve resilience and optimize system operation	
		Implementing conservation measures to reduce demand	
		Using city's reservoirs as flood control structures	
	Decreased quality of water supply	Filtration avoidance determinations: land acquisition, land use, forest management	
		Increasing flow to filtered systems during high turbidity periods after heavy precipitation	
		Treating water in aqueducts with alum to coagulate particulates after a heavy storm event	
	Increased demand	Incorporating climate change adaptive measures into new projects	
		Imposing and enforcing water conservation measures	
		Improving the enforcement of illegal fire hydrant use	
		Increasing the number of public swimming pools	

Source: New York City Department of Environmental Quality

<table>
<tr><th>City/Utility</th><th>Climate Change Impacts to Water Supply</th><th>Adaptive Measures in Place</th><th>Year CAP Created</th></tr>
<tr><td>City and County of Denver</td><td>Decreased quantity of water supply (reduced snowpack and earlier snowmelt)</td><td>Implementing water conservation programs: rebates for replacing bathroom and kitchen fixtures and toilet with low-flow alternatives; encouraging customers to install irrigation systems scheduled to real-time weather conditions; creating a recycled water system</td><td rowspan="4">2014</td></tr>
<tr><td rowspan="3"><u>Population Served:</u> 1.3 million
<u>Main Water Source:</u> Surface
<u>Budget Estimate:</u> $33 million (for the implementation of both watershed protection projects and forest treatment)</td><td rowspan="3">Decreased quality of water</td><td>Improving stormwater infrastructure to protect rivers</td></tr>
<tr><td>Developing Water Quality Strategic Plan to consider increasing water quality treatment</td></tr>
<tr><td>Increasing water quality monitoring in watersheds</td></tr>
</table>

Source: Denver Environmental Health

<table>
<tr><th>City/Utility</th><th>Climate Change Impacts to Water Supply</th><th>Adaptive Measures in Place</th><th>Year CAP Created</th></tr>
<tr><td>City of Chicago</td><td rowspan="3">Decrease in water quality (increase in concentrations of waterborne pollutants due to increased evaporation from lake surfaces; increase in sewage and pollutants entering waters due to flooding and sewer overflows)</td><td>Implementing a stormwater management ordinance to increase permeable area in new development projects</td><td rowspan="3">2008</td></tr>
<tr><td rowspan="2"><u>Population Served:</u> 2.7 million
<u>Main Water Source:</u> Surface
<u>Budget Estimate:</u> $142 million received in funding thus far</td><td>Installing green alleys to create more permeable surfaces and increase groundwater infiltration</td></tr>
<tr><td>Developing a sewer model to assess vulnerable areas that require capital improvements or green infrastructure</td></tr>
</table>

Source: Climate Adaptation Knowledge Exchange (CAKE).

<table>
<tr><th>City/Utility</th><th>Climate Change Impacts to Water Supply</th><th>Adaptive Measures in Place</th><th>Year CAP Created</th></tr>
<tr><td>San Francisco Public Utilities Commission/San Francisco Department for the Environment</td><td>Sea level rise causing saltwater intrusion</td><td>Altering groundwater pumping strategies; in-lieu recharge to increase groundwater storage</td><td rowspan="2">2004</td></tr>
<tr><td><u>Population Served:</u> 2.6 million
<u>Main Water Source:</u> Surface
<u>Budget Estimate:</u> Not available</td><td>Decrease in water availability for summer months as warmer temperatures reduce snowpack which acts as stored water for summer months</td><td>Diversifying water supply by supplementing surface water with groundwater and utilizing recycled water for irrigation</td></tr>
</table>

Source: San Francisco Department of Environment and San Francisco Public Utilities Commission (SFPUC).

USEPA Sustainable Water Infrastructure Program

USEPA provides federal and state funding specific to water and wastewater utilities for the purpose of increasing resilience. These guides and grants include projects that target water and energy efficiency, green infrastructure, water reuse, and flood protection (USEPA).

NOAA Sea Grant Coastal Communities Climate Adaptation Initiative

This competition was designed to help coastal communities prepare for the effects of climate variability and change. Proposed projects must address the potential climate change impacts that the coastal community faces (NOAA).

Federal Emergency Management Agency (FEMA) Hazard Mitigation Grant Program

Grants are only administered through this program following a presidential major disaster declaration. Funding can be used to implement mitigation projects that will reduce the loss of human life and damage to infrastructure as a result of natural disasters. To qualify, potential savings must outweigh the cost to implement the project (FEMA).

FEMA Pre-disaster Mitigation Grant Program

This program targets preemptive mitigation strategies. Eligible projects will be designed to increase the resilience of applicable structures and therefore reduce the risk to human life in the event of a natural disaster (FEMA).

United States Department of Agriculture (USDA) and Rural Development Emergency Community Water Assistance Grants

These grants are intended for emergency preparedness and recovery. Emergencies include extreme weather events resulting from climate change, such as increased frequency and intensity of flooding, droughts, and hurricanes. Funds can also be applied to emergencies such as chemical spills or earthquakes. To qualify, the applicant must operate in rural tribal lands, colonies, or a town with 10,000 or fewer people. Per the US Department of Agriculture, the area's median household income must also be less than the state's median household income for non-metropolitan areas.

REFERENCES

Climate Adaptation Knowledge Exchange (CAKE). A Roadmap for Action: The Chicago Climate Action Plan. 2013. Updated 2020. https://www.cakex.org/case-studies/roadmap-action-chicago-climate-action-plan (Accessed Jun 3, 2020).

Denver Environmental Health. City and County of Denver: Climate Action Plan. 2014. https://www.denvergov.org/content/dam/denvergov/Portals/771/documents/EQ/Climate1/Climate_Adaptation_Final%20with%20letter.pdf (Accessed June 3, 2020).

NACWA (National Association of Clean Water Agencies) and AMWA (Association of Metropolitan Water Agencies). 2010. Confronting Climate Change: An Early Analysis of the Cost of Climate Adaptation. Englewood, Colo.: CH2M Hill. https://www.amwa.net/assets/ConfrontingClimateChangeOct09.pdf (accessed Oct. 6, 2020).

New York City Department of Environmental Protection. *The NYCDEP Climate Change Program: Assessment and action Plan*. 2008. https://data.globalchange.gov/report/nycdepclimchprogram-assessment-2008 (accessed June 3, 2020).

San Francisco Department of Environment and San Francisco Public Utilities Commission (SFPUC). 2004. *Climate Action Plan for San Francisco: Local Actions to Reduce Greenhouse Gas Emissions.* http://www.sfenvironment.org/sites/default/files/fliers/files/climateactionplan.pdf (accessed June 3, 2020).

Snover, A.K., Whitely Binder, L.C., Lopez, J., Willmott, E., Kay, J.E., Howell, D. and Simmonds, J. 2007. Preparing for Climate Change: A Guidebook for Local, Regional, and State Governments. University of Washington Climate Impacts Group and King County, Washington, in association with and published by ICLEI – Local Governments for Sustainability, Oakland, Calif.

USEPA (US Environmental Protection Agency). n.d. Climate Resilience Evaluation and Awareness (CREAT) Climate Scenarios Projector. http://www2.epa.gov/crwu/view-your-water-utilitys-climate-projection-scenario-based-projected-changes-map (accessed June 3, 2020).

USEPA. 2015. Adaptation Strategies Guide for Water Utilities (pp. 47-66). https://www.epa.gov/sites/production/files/2015-04/documents/updated_adaptation_strategies_guide_for_water_utilities.pdf (accessed Oct. 6, 2020).

USEPA. 2011. Climate Change Vulnerability Assessments: Four Case Studies of Water Utility Practices. Washington, DC, EPA/600/R-10/077F, 2011.

US Global Change Research Program. 2014. *Climate Change Impacts in the United States.* Appendix 5: Scenarios and Models (pp. 821-826). http://s3.amazonaws.com/nca2014/high/NCA3_Climate_Change_Impacts_in_the_United%20States_HighRes.pdf (Accessed June 3, 2020).

Water Research Foundation. *Effects of Climate Change on Water Utilities.* 2008. https://www.waterrf.org/system/files/resource/2019-05/EffectsofClimateChangeOnWaterUtilities.pdf (Accessed June 3, 2020).

This page intentionally blank.

Chapter 10

Case Studies of Climate Action Plan and Adaptation Measures

In an effort to conduct a gap analysis and perform a review of the information that is available in other reports generated with the purpose similar to that of this manual, the authors of this manual conducted a thorough review of various reports published by other sources. The purpose of the review was to identify any similarities between the stated goals for developing this manual and other reports available in the public sphere to avoid repetition and reproduction of the work already done by other researchers in the climate change community. The following is a summary of the latest sites and literature that were reviewed in an effort to prevent reproducing resources already out there and provide an approach for developing a water utility climate action plan (CAP) that considers both mitigation (e.g., greenhouse gas reductions) and adaptation.

As a primary source of information, the US Environmental Protection Agency (USEPA) has a great site of resources. This site provides a list of existing resources that address climate change adaptation planning—including tools, guidebooks, clearinghouses, and other resources from states, federal agencies, nonprofit and private organizations, and others (USEPA). Following are some commonly referenced documents that could be used for developing various portions of our handbook, as well as to provide guidance on CAPs.

1) Good outline for climate change action plan support is summarized by the International Council for Local Environmental Initiatives (ICLEI) and others:

Preparing for Climate Change: A Guidebook for Local, Regional, and State Governments

2) California Department of Water Resources: *Climate Change Handbook for Regional Water Planning*
3) USEPA: *Adaptation Strategies Guide for Water Utilities*
4) The Water Resources Adaptation to Climate Change Workgroup of the Advisory Committee on Water Information's report, *Next Steps for Managing Freshwater Resources in a Changing Climate,* released April 2014. The report summarizes recommendations for next steps in implementing the *2009 National Action Plan: Priorities for Managing Freshwater Resources in a Changing Climate.*

Included in the next sections is a summary of the review conducted by the manual authors for the various reports developed by the climate change scientific community to provide guidance on the climate change adaptation measures and CAP preparation. The review was conducted by posing three questions:

- What is the approach for CAP used in this study?
- Are there elements of this case study that we can add to our manual content as useful information for others?
- Are there items in the case study we are regurgitating or repeating in the manual?

CASE STUDIES

Mitigation and Adaptation Efforts in EPA Region 3

Below is a review of the study Climate Change Implementation Plan prepared by the Region III Climate Network Climate Adaptation Working Group at USEPA Mid-Atlantic Region III.

1. What is the approach for CAP used in this study?

This document summarized climate change mitigation and adaptation measures taken by 13 utilities in the Northeast United States (USEPA Region 3). The CAPs are focused on documenting the issues faced by the utilities, the threats and risk to the utilities because of the issues, and the steps taken to address the issues. Each report also includes the cost of the adaptation and mitigation measure, the cost benefit of taking the measure, and the source of funding for the project. Most examples are associated with water and wastewater utilities.

2. Are there elements of this case study that we can add to our manual content as useful information for others?

This document is a good summary of effective case studies for utilities that have addressed the issue of climate change impacts. The discussion on cost of the adaptation or mitigation measure, the benefit (cost, perception, and environmental), and the source of funding are helpful and should be incorporated to the manual content. The document also includes a list of common threats to small utilities due to climate change. These common threats should be captured in the focus areas chapter. Each utility is different and so is the impact of climate change to the utility. The solution in the form of adaptation and mitigation measure is also very specific to the utility. This should be included into the discussion.

3. Are there items in the case study we are regurgitating or repeating in the manual?

No. In fact, the case studies in the report include a discussion of the problems faced by the utility if an adaptation or mitigation measure is not taken. The quantification of the value lost because of inaction should be incorporated into the discussion of the CAP.

Vulnerability Assessment Approaches Taken by US Water Utilities

Below is a review of the study conducted by USEPA evaluating Climate Change Vulnerability Assessment: A Review of Water Utility Practices.

1. What is the approach for CAP used in this study?

This document summarized climate vulnerability assessment of eight water utilities. The CAP is focused toward mitigation.

2. Are there elements of this case study that we can add to our manual content as useful information for others?

The climate change vulnerability assessment approach used in both drinking water and wastewater utilities for the risk assessment can be useful for the manual. The top-down modeling assessment and the bottom-up threshold analysis approach described in the case studies can be helpful for many utilities to provide guidance on climate change vulnerability assessment studies.

3. Are there items in the case studies we are regurgitating or repeating in the manual?

The approach described in the case studies are top-down assessment and bottom-up climate change assessment. The top-down approach, which is the most popular approach used by most of the utilities, is model and data-driven, and more time- and resource-intensive than bottom-up assessment. In general, the top-down assessment requires expertise whereas the bottom-up assessment is driven by knowledge of the utility system itself. Case studies of these approaches from both drinking water and wastewater utilities can be useful examples for the manual.

City of Chula Vista (San Diego County)—Climate Action Planning—Climate Change Working Group

Below is a review of the study conducted by City of Chula Vista in San Diego County summarizing the climate action plan for the City.

1. What is the approach for CAP used in this study?

The City of Chula Vista's CAP looks at both mitigation and adaptation measures for the city as a whole. Mitigation in the city's CAP refers to reducing greenhouse gas (GHG) emissions (focused on carbon dioxide at first but has expanded to additional GHG emissions). The original report in 2000 was a standalone, mitigation-focused report and provided the approach for developing a baseline of CO_2 emissions, setting a reduction goal (which is based on the United Nations Framework Convention on Climate Change, the ICLEI Cities for Climate Protection campaign and the US Conference of Mayor's Climate Protection Agreement, at a local governmental level to fit within international climate treaty reduction targets) and identifying policy measures that will achieve those reductions. Water measures were focused on conservation.

In 2010, the city evaluated the potential impacts from climate change on municipal infrastructure and services and recommended strategies to adapt to those potential impacts. The city's CAP recommended 11 strategies to adapt the community to these impacts within energy and water supply, public health, wildfires, ecosystem management, coastal infrastructure, and the local economy sectors. Strategy #4 is focused on developing local water supply and reuse, and Strategy #5 is focused on stormwater pollution prevention and reuse.

2. Are there elements of this case study that we can add to our manual content as useful information for others?

Mitigation: We can reference the methodologies used for emissions estimating and reasoning for goal setting (if not already mandatory or if the city has not already committed voluntarily to a goal).

Adaptation: We can summarize the city's approach to improving water supply resilience through investigating ways of developing a more local, sustainable supply of water through building codes, incentives, and best practices. Each report provides metrics and progress reports, which are an important piece of the approach that we need to highlight.

3. Are there items in the case study we are regurgitating or repeating in the manual?

Not likely. These CAPs are aimed at a municipal local government and are not specific to the water sector. The discussion includes the water sector, but very briefly in the mitigation report and in a couple city-specific strategies in the adaptation report. It will be good to glean specific strategies from the adaptation report.

Niagara Region Community Climate Change Action Plan—Partners for Climate Protection: Milestones 2 and 3

Below is a summary of the study performed by Niagara Region Community Climate Change Action Plan.

1. What is the approach for CAP used in this study?

Addresses mostly mitigation and only cursory mention of adaptation, following the five milestones of the Federation of Canadian Municipalities' Partners for Climate Protection (PCP) Program. This program outlines five milestones for municipalities to work through to act on climate change, both corporately and by the community. The document is a very high-level summary of goals and accomplishments.

2. Are there elements of this case study that we can add to our manual content as useful information for others?

Mitigation: Yes, but largely focused on sectors other than water, such as agriculture, building stock, industry.

Adaptation: Makes very little mention of adaptation other than to reference a separate study, "Adapting to Climate Change: Challenges for Niagara," authored by Dr. Jennifer Penney, which is available online. (Penney, 2012). The report provides a useful overview of climate impacts and adaptation strategies.

3. Are there items in the case study we are regurgitating or repeating in the manual?

No, this case study does not provide much in the way of specific guidance, other than referencing the guidelines of the PCP and the Engineers Canada infrastructure risk assessment methods.

A Climate Action Plan for Waterloo Region – Living Smarter in 2020

Below is the review of the study performed by the City of Waterloo in Canada summarizing the climate action plan for the region and actionable items for living smarter in 2020.

1. What is the approach for CAP used in this study?

The focus of this CAP in entirely on mitigation plan; there is nothing addressing adaptation. It lays out the need for emissions reduction from a global and local perspective, and provides broad discussion of the benefits of such reductions from health, environmental sustainability and economic perspectives. A goal is set right up front: reduce carbon emissions from 2010 levels by 6% by 2020. The CAP appears to have been guided by Canada's Partners for Climate Protection (PCP) program, which includes over 240 municipalities that have committed to reducing GHGs and acting on climate change. The PCP lays out five milestones:

1) Create a community emissions inventory and forecast.
2) Set community emissions reduction targets.
3) Develop a community climate action plan (CAP).
4) Implement community CAP.
5) Monitor progress and report results.

It is municipality and community focused (i.e. not utility focused), and emphasizes collaborative reduction of GHG by municipal government and community leaders, acknowledging that in Canada municipalities have direct control or indirect influence over 44% of all GHG emissions in Canada. It includes a comprehensive inventory of 3.6 MT of annual GHG emissions sources for the Waterloo Region with do nothing projections:

1) Transportation – 40%
2) Workplaces 32%
3) Homes 22%
4) Agriculture & Food – 5%
5) Waste – 1%

Each of these sectors is broken down further, and areas of potential significant GHG reduction are identified, followed by specific opportunities and programs for GHG reductions:

1) Areas of focus:
    - Building construction and energy usage
    - Local land use
    - Agricultural practices
    - Transportation patterns
    - Economic development

2) Key solutions for achieving GHG reduction goals include:
   - Education programs across all sectors with more collaboration among stakeholders
   - Energy rebate programs for homeowners and commercial buildings
   - Building code changes
   - Revised eco-friendly master plans for transportation, zoning, etc.
   - Tap into national funding programs
   - Adopting new technologies and energy concepts and green energy solutions

2. Are there elements of this case study that we can add to our manual content as useful information for others?

Given the broader focus on community actions as opposed to utilities, opportunities to add to our manual content are limited. However, it does provide some good stakeholder engagement steps and includes a fairly comprehensive assessment of GHG inventory for establishing a baseline from which to measure progress. If we are to include guidance on developing baseline and sources for GHG emissions, some of this content may be helpful.

It also emphasizes the interconnectedness with existing policies, programs, and activities led by the community and government. This may be an important concept to emphasize in our manual where utilities may have influence; for example, by influencing plumbing codes to adopt conservation fixtures or simply assessing and adjusting our standard utility practices for materials, fleet, workflow, etc. to consider GHG impacts.

3. Are there items in the case study we are regurgitating or repeating in the manual?

There are a few areas of crossover, such as stakeholder engagement, assessment of energy audit/inventory of GHG emissions, and incentive programs; however, given the narrower focus on mitigation only and the limited acknowledgement of water and wastewater utility opportunities, crossover areas are limited.

Kelowna's Community Climate Action Plan

Below is a summary of the climate action plan developed by the City of Kelowna's community.

1. What is the approach for CAP used in this study?

The CAP is primarily focused on climate mitigation. That is to say, reducing GHG emissions (primarily carbon dioxide but not exclusively). The approach focuses on identifying the baseline of GHG emissions, setting a goal for reduction (which is derived from a political process at a higher governmental level), and finally identifying a portfolio of options that will achieve those reductions. Although there is brief mention of adaptation (less than one page out of 70, total), the CAP is essentially focused exclusively on mitigation.

2. Are there elements of this case study that we can add to our manual content as useful information for others?

Possibly, but it may require additional research. There is considerable data contained within about the sources of emissions within the city/region. Some of the methodology to identify emissions may be useful for utility carbon emissions management. The portfolio approach used (after setting the goal, determining where the reductions will come from in a clear manner) may also be useful. There is nothing in this CAP that would be useful on the adaptation side.

3. Are there items in the case study we are regurgitating or repeating in the manual?

Not very likely. This study is aimed at a whole municipal area and is not at all specific to the water sector. The discussion includes both municipal solid waste (as a major, but nowhere near the largest, mitigation option) and the water sector, but the water sector is only mentioned briefly in passing in the very short adaptation section. Anything with sector-specific actions will not be a repeat of information contained in this CAP.

Metro Vancouver Sustainable Region Initiative: Corporate Climate Action Plan

Summarized below is a review of the study conducted by Metro Vancouver. The study includes a Corporate Climate Action Plan that was developed as part of the Sustainable Region Initiative.

1. What is the approach for CAP used in this study?

The approach is largely focused on mitigation and further focused on energy use/recovery. Some discussion on sequestration. Some discussion on adaptation in terms of adapting existing operations and building resilient new infrastructure.

- Power reliability (i.e., increased risk of outages)
- Retrofitting infrastructure at risk of flooding (building elevations, dykes
- Securing property for future adaptive requirements (more dykes, more peaking reservoirs, etc.)
- Increased storage during wet season/reducing demand during dry season

2. Are there elements of this case study that we can add to our manual content as useful information for others?

The whole document is fewer than 20 pages, just outlining major principles. This is something to consider: that a CAP does not necessarily have to be long and analytical. Major principles are sound and should be mentioned, but these major principles would have been mentioned anyway.

3. Are there items in the case study we are regurgitating or repeating in the manual?

There may be some repetition of basic principles.

Flood Resilience Pilot Project, Berwick Water Department, Berwick, Maine

This resulted in the recent USEPA release of "Flood Resilience: A Basic Guide for Water and Wastewater Utilities" (USEPA 2014).

1. What is the approach for CAP used in this study?

The focus was on assessing and managing flooding risks for medium and small scale water and wastewater facilities. The document's approach embraces that the frequency and magnitude of flood impacts are being affected by climate change, so additional adaptation planning is required. The document offers a four-stage resiliency planning framework that assesses flood dynamics, critical vulnerabilities, risk reduction options, and implementation plans (prioritizing) for risk reduction measures.

2. Are there elements of this case study that we can add to our manual content as useful information for others?

The guide's steps are valid. There are worksheets within this document that at times are very clear and helpful. The best elements of these worksheets could be adapted to a broader scope for initial climate risk assessment of water infrastructure. Worksheets are provided in sections where they offer clear, guiding questions that lead the user to proceed through reasonable steps in a planning process. Some good questions are also provided in sidebar content. Accompanying videos are provided for additional worksheet support.

3. Are there items in the case study we are regurgitating or repeating in the manual?

Only in a good way. This document hits some key areas that our manual would want to emphasize, including the following.

- Climate Risk/Adaptation Planning involves iterative cycles, in a structured process.
- Levels of service are important to early in your planning.
- Vulnerability assessment for equipment, processes and people/roles should be conducted.
- It is important to prioritize adaptation measures.

REFERENCES

Penney, J. 2012. *Adapting to Climate Change: Challenges for Niagara.* 2012. http://www.niagararegion.ca/government/planning/pdf/climatechangerport.pdf. (accessed June 3, 2020).

USEPA (US Environmental Protection Agency). 2014. *Flood Resilience: A Basic Guide for Water and Wastewater Utilities.* https://www.epa.gov/sites/production/files/2015-08/documents/flood_resilience_guide.pdf (accessed June 3, 2020).

Chapter 11

Summary and Conclusions

There is no bigger concern for utility managers than that of not providing the services intended to be provided by the utility. While many factors impact a utility's ability to serve its customers' needs, no other factor impacts the utility operations in a complex, uncertain, and overwhelming way like climate change does. There are already several instances of utilities facing the worst-case impacts driven by climate change and being caught unprepared and off-guard in responding to the impacts of climate change. The impacts of various hurricanes on the East and Southeast coasts of the United States are a good example. The staggering impact of the wildfires in California is another example of the overwhelming scale and magnitude of the climate change impact. In the face of uncertainty and the exponential increase in the frequency of extreme events, it seems prudent for utilities to understand the impacts of climate change and their system vulnerabilities and develop climate action plans (CAPs) to stay prepared and resilient.

Until recently, modeling climate change and quantifying its impacts were less known fields of science. There have been drastic improvements in the human understanding of the mechanics of climate change and how it impacts each and every region. Historically, a utility's planning and management were primarily based on the assumption of a repeat of a historical worst-case scenario. Today, we know that planning for historical worst-case scenarios leaves us underprepared for the future. Not only do we know there is impact associated with climate change, we also know and can meaningfully quantify the inherent uncertainty associated with the phenomenon. With this knowledge, there is no better time for developing and maintaining CAPs than now. Many climate change experts in the scientific community have taken this issue on, and the byproduct of that is the wealth of information available to utility managers as they prepare to build resiliency into their operations. However, awareness of the available resources and getting access to them may sometimes be an issue, even if one does not know where to look.

The primary objective of preparing this manual was to provide a hands-on guiding document to utility managers as they tackle the issue of creating and implementing CAPs. The manuscript was prepared with the intention of providing support to small and medium utility managers as they try to incorporate climate change effects into their utility planning and operations processes. Most of the focus for the document was limited to water utilities, and less focus on wastewater utilities. While responses to climate change impacts can be in the form of adaptation and mitigation measures, most of the focus for this manual was on the adaptation measures. The manual walks the reader through the logical process of introducing the problem, explaining the climate change science to a relevant degree, and describing engineering practices in pre–climate change and post–climate change awareness environments. The manual then outlines what a CAP does and does not include; it is helpful to define the features of the CAP to understand the level of effort required to prepare one. Then the discussion is focused on why utilities should care about preparing CAPs and the issues associated with lack of preparation and a no-action approach. The manual includes a detailed summary of the components of a CAP and the relevant resources available to develop them. The manual also includes a meaningful discussion on public participation and stakeholder engagement. Utility managers know that getting a rate change approved by their rate payers and stakeholders is an uphill battle of its own. Getting rate payers and stakeholders onboard with controversial and complex topics like climate change and CAPs can be a challenging task. However, there is much guidance available from both experts in the water sector and other utilities that have been through this process in the past. The manual includes a discussion on the cost associated with preparing CAPs and various sources for funding CAPs. A useful summary of case studies that are relevant to the utility's efforts are included in the discussion for the benefit of the utility managers.

This manual was an earnest attempt by the authors to develop a resource material that is well founded in scientific expertise and one that provides benefit to one and all in the engineering community. Innovation and efficiency are more easily achieved when we share our experience with others in similar conditions and also when we learn from others. The objective of this manual is served if it initiates a dialogue within individual utilities, helps utility managers to realize the importance of CAPs, and gets them to the starting point of preparing a CAP. The ultimate goal of this manual is to help utilities become resilient to climate change impacts, one utility at a time, thus resulting in an overall systemic resiliency in the water infrastructure all over the United States and the throughout the world.

Glossary

adaptation: actions and responses that may be taken to address potential effects from climate change (e.g., new water treatment and conveyance designs, enhanced stormwater management practices)

adaptation measures: adopting innovative ways to achieve goals

adaptive capacity (NRC, 2010a-d): the ability of a system to adjust to climate change (including climate variability and extremes) to moderate potential damages, to take advantage of opportunities, or to cope with the consequences

American Metropolitan Water Associations (AMWA): a nonprofit organization serving the water profession in the United States

American Water Works Association (AWWA): a nonprofit organization serving the water profession in North America

Anthropogenic: pertaining to or involving the impact of humans

aquifer: a geological formation, group of formations, or part of a formation that is saturated and sufficiently permeable to transmit quantities of water to wells and springs

atmosphere: the gaseous envelope (air) surrounding the earth. It consists of oxygen (O_2), nitrogen (N_2), and other gases, extends to a height of about 22,000 miles (35,400 km), and rotates with the earth

bell-curve distribution, also called normal distribution: half of the data will fall to the left of the mean, and half of the data will fall to the right of the mean

Bermuda High: the Bermuda High situated over the North Atlantic Ocean is associated with the regional climate over the eastern United States. The Bermuda High also defines the low-level jet of southerly flows over the Southern Great Plains

best management practice: a practice, measure, or procedure that is beneficial, empirically proven, cost-effective, and widely accepted by the professional community

carbon footprint: the total greenhouse gas or carbon dioxide emissions by an individual, event, group, etc.

carbon tax: a fee imposed on the burning of carbon-based fuels such as coal, oil, or gas

cap-and-trade: regulatory program designed to limit the total emission of carbon dioxide and other chemicals produced by industrial activities.

climate: regional weather conditions averaged across a period of years, or the statistical distribution of weather over the long term, typically interpreted as at least 30 years

climate action plan (CAP): a plan that prepares a water/wastewater system for the effects of climate change

climate change: a change in the statistical distribution—the state of climate—over decades

climate model: a numerical representation of the climate system based on the physical, chemical, and biological properties of its components, their interactions, and feedback processes

codes and standards: a set of requirements governing the design and performance of water-using fixtures or devices adopted by a nationally or internationally recognized code-setting or standard-setting organization

condensation: conversion from the vapor state to the liquid state. A common example is the formation of water droplets on the outside of windows in humid climates

demand management: a strategic practices that optimizes water supply, treatment, and delivery requirements to support long-term sustainability of water resources

dendrohydrology: the science that uses past events or climatic events by studying the growth of tree rings to investigate and reconstruct hydrologic processes such as river flow

desalination (*also referred to as desalting*): a water treatment process, such as distillation, reverse osmosis, or electrodialysis, that removes dissolved mineral salts and other dissolved solids from water

direct potable reuse: highly treated reclaimed water as part of the potable water supplied directly to a water treatment plant or distribution system without the use of an environmental buffer and with or without retention in an engineered storage buffer

drought: a normal recurrent feature of climate that occurs in virtually all climate zones. Drought occurs primarily because of a deficiency of precipitation over an extended period of time resulting in a water shortage

drought management: actions and policies that alter the management of water supplies and customer water demands to mitigate the impacts of drought, e.g. implementation of conservation measures

emission: a venting of gas or radiation from a building or piece of equipment

evaporation: a process in which a liquid is changed by volatilizing (boiling) to a gaseous state at a set of temperature and pressure conditions

freshwater lens: a layer of fresh groundwater that typically floats on top of saltwater and is recharged by precipitation. Rising sea level diminishes the area above the sea level where the lens can reside

green infrastructure: an approach to water management that uses either natural or engineered systems that mimic natural processes to control runoff and reduce runoff demand

hydrologic cycle: the natural process of recycling water from the atmosphere down to (and through) the earth and back to the atmosphere

integrated resource management: the use of systematic procedures to allocate and use in terms of timing and mix, all available assets to meet the objectives of a company or organization

integrated resources planning: a comprehensive form of water utility planning that encompasses least-cost analysis of demand management and supply management options, as well as an open and participatory decision-making process, preparation of alternative planning scenarios, and the competing policy goals among them

indirect reuse: the use of reclaimed water for nonpotable or potable purposes by first discharging to a water supply source, such as fresh surface water or groundwater, where it mixes, dilutes, and may be transformed before it is removed for reuse

Kyoto Protocol: adopted in 1977, this protocol commits industrialized countries to limit and reduce greenhouse gas (GHG) emissions in accordance with agreed individual targets

mitigation: the act or process of avoiding, minimizing, rectifying, reducing, and compensating for impacts; to moderate or make less severe

National Association of Clean Water Agencies (NACWA): a nonprofit organization serving the wastewater profession in the United States

nonpotable: not safe to drink

ocean acidification: when atmospheric CO_2 is absorbed by seawater and the water becomes more acidic

paleohydrology: the study of the evidence of the movement of water before recordings or historical observation

paleoclimate: a climate existing at a particular time in the geological past

potable: safe to drink

precipitation: (1) the total measurable supply of water received directly from clouds as rain, hail, mist, sleet, or snow, usually expressed as depth in a day, month, or year (2) the process by which atmospheric moisture is discharged onto a land or water surface

probabilistic: not able to predict future events definitively

quantitative: pertaining to size, magnitude, or degree

reclaimed water: wastewater that has been treated and recovered for useful purposes, such as landscape irrigation

recycled water: water that is reused in the same process, with or without additional treatment, or reused by the same user

reuse: the application of appropriately treated wastewater to a constructive purpose

resilience: the ability of a system (e.g., ecosystem) to recover from or resist stress

risk: (1) the probability that a particular adverse event will occur (2) the measure of the potential damage to, or loss of, an asset based on the probability of an undesirable event

risk management: the process of weighing policy alternatives, selecting the most appropriate regulatory action, and integrating the results of a risk assessment

salt lens: area where saltwater resides; sea level can push salty water upstream in coastal areas, threatening surface water supplies

sea breeze: a coastal phenomenon that influences meteorological conditions at the local scale

snowpack: a field of naturally packed snow that provides a steady supply of water and is also useful in furnishing water power. This term is used in the mountain regions of the United States. Snowpack is also called snow storage or, sometimes, snow cover.

stationarity: the idea that natural systems fluctuate within an unchanging envelope of variability

storm surge: a rise or accumulation of water against the shore produced by wind stress and atmospheric pressure differences in a storm

surface water supply: water supplied from a stream, lake, or reservoir

vulnerability: the susceptibility or likelihood that a public water system or source water would be contaminated or otherwise adversely affected by an activity, practice, or unexpected emergency situation

water intake: (1) the total gross quantity of water withdrawn from supply in a given interval, usually a day, month, or year (2) the physical structure capable of removing water from a source, such as a lake

water losses: the difference between system input volume and authorized consumption, consisting of apparent losses plus real losses

Water Research Foundation (WRF): a nonprofit organization that sponsors research for the water profession (formerly AwwaRF)

water resources management: the monitoring, planning, and administration of water resources for various purposes

WaterSense: a voluntary partnership program sponsored the US Environmental Protection Agency (USEPA) that provides guidance for water-efficient products and is a resource for water conservation

This page intentionally blank.

Acronyms

AMO	Atlantic Multidecadal Oscillation
AOGCM (GMC)	atmosphere–ocean general circulation model
BCP	business continuity plan
CAKE	Climate Adaptation Knowledge Exchange
CAP	climate action plan
CCP	crisis communications plan
CEMP	comprehensive emergency management plan
CEQA	California Environmental Quality Act
CIG	Climate Impact Group (University of Washington)
CMIP3	Coupled Model Intercomparison Project 3
COOP	continuity of operations plan
CREAT	Climate Resilience Evaluation and Awareness Tool (USEPA)
CSCs	US Department of Interior Climate Science Centers
CSO	combined sewer overflow
CUWA	California Urban Water Agencies
DBP	disinfection byproduct
EAP	emergency action plan (also known as ERP)
ENSO	El Niño–Southern Oscillation
EOP	emergency operations plan
ERP	emergency response plan (also known as EAP)
FHWA	Federal Highway Administration
GCM	general circulation model
GHG	greenhouse gas
IOD	Indian Ocean Dipole
IDF	intensity-duration-frequency
IPCC	United Nations Intergovernmental Panel on Climate Change
IPCC AR4	IPCC Fourth Assessment Report
IPCC TAR	IPCC Third Assessment Report
IRP	integrated resource planning
LOS	level of service
MJO	Madden-Julian Oscillation
NACWA	National Association of Clean Water Agencies
NAO	North Atlantic Oscillation
NOAA	National Oceanic and Atmospheric Administration

NPO	North Pacific Oscillation
NPGO	North Pacific Gyre Oscillation
O&M	operations and maintenance
PDO	Pacific Decadal Oscillation
PIEVC	Public Infrastructure Engineering Vulnerability Committee
RCM	regional climate model
RCP	Representative Concentration Pathway
RISA	Regional Integrated Sciences and Assessments
SCADA	supervisory control and data acquisition
SNOTEL	snow telemetry
SRES	Special Report on Emissions Scenarios
SSO	sanitary sewer overflow
TOC	total organic carbon
VA	vulnerability assessment
WBCSD	World Business Council on Sustainable Development
WRI	World Resources Institute
WUCA	Water Utility Climate Alliance
WWTP	wastewater treatment plant

Index

Note: *f.* indicates figure; *t.* indicates table

O

P

Q

R

S

T

U

V

W

AWWA Manuals

M65, *On-Site Generation of Hypochlorite,* #30065

M66, *Cylinder and Vane Actuators and Controls—Design and Installation,* #30066

M68, *Water Quality in Distribution Systems,* #30068

M69, *Inland Desalination and Concentrate Management,* #30069

M77, *Condition Assessment of Water Mains,* #30077

www.ingramcontent.com/pod-product-compliance
Lightning Source LLC
LaVergne TN
LVHW082341200926
848481LV00012B/3412